PhysioEx™ for Human Physiology

SECOND EDITION

TIMOTHY STABLER
Indiana University Northwest

MARCIA C. GIBSON
University of Wisconsin—Madison
(Histology Review Supplement)

Benjamin
Cummings

San Francisco   Boston   New York
Cape Town   Hong Kong   London   Madrid   Mexico City
Montreal   Munich   Paris   Singapore   Sidney   Tokyo   Toronto

Publisher: Daryl Fox

Project Editor: Barbara Yien

Managing Editor: Wendy Earl

Production Supervisor: Janet Vail

Benjamin Cummings gratefully acknowledges Carolina
Biological Supply for the use of numerous histology
images found on the PhysioEx CD-ROM.

PhysioEx™ 4.0 Products

For Anatomy & Physiology

PhysioEx™ 4.0
Laboratory Simulations in Physiology (CD Version):
order ISBN 0-8053-6144-8

PhysioEx™ 4.0
Laboratory Simulations in Physiology (Web Version):
order ISBN 0-8053-6150-2

Instructor's Guide for PhysioEx™ 4.0:
order ISBN 0-8053-6155-3

For Human Physiology

PhysioEx™ for Human Physiology 2nd ed. (CD Version):
order ISBN 0-8053-6153-7

PhysioEx™ for Human Physiology 2nd ed. (Web Version):
order ISBN 0-8053-6151-0

Instructor's Guide for PhysioEx™ for Human Physiology
2nd ed.: order ISBN 0-8053-6154-5

To locate the Benjamin Cummings sales representative
nearest you, visit **http://www.aw.com/replocator**

Benjamin
Cummings

ISBN 0-8053-6152-9

3 4 5 6 7 8 9 10—MAL—06 05 04 03 02

www.aw.com/bc

CONTENTS

PREFACE

PhysioEx™ for Human Physiology 2nd ed. consists of ten physiology lab simulations that may be used to supplement or replace wet labs. This easy-to-use software allows you to repeat labs as often as you like, perform experiments without harming live animals, and conduct experiments that may be difficult to perform in a wet lab environment due to time, cost, or safety concerns. You also have the flexibility to change the parameters of an experiment and observe how outcomes are affected. In addition, PhysioEx includes an extensive histology tutorial that allows you to study histology images at various magnifications. This manual will walk you through each lab step-by-step. You will also find Review Sheets in the back of your manual to test your understanding of the key concepts in each lab.

Topics

Exercise 1 Cell Transport Mechanisms and Permeability. Explores how substances cross the cell's membrane. Simple and facilitated diffusion, osmosis, filtration, and active transport are covered.

Exercise 2 Skeletal Muscle Physiology. Provides insights into the complex physiology of skeletal muscle. Electrical stimulation, isometric contractions, and isotonic contractions are investigated.

Exercise 3 Neurophysiology of Nerve Impulses. Investigates stimuli that elicit action potentials, stimuli that inhibit action potentials, and factors affecting nerve conduction velocity.

Exercise 4 Endocrine System Physiology. Investigates the relationship between hormones and metabolism; the effect of estrogen replacement therapy; and the effects of insulin on diabetes.

Exercise 5 Cardiovascular Dynamics. Topics of inquiry include vessel resistance and pump (heart) mechanics.

Exercise 6 Cardiovascular Physiology. Variables influencing heart activity are examined. Topics include setting up and recording baseline heart activity, the refractory period cardiac muscle, and an investigation of physical and chemical factors that affect enzyme activity.

Exercise 7 Respiratory System Mechanics. Investigates physical and chemical aspects of pulmonary function. Students collect data simulating normal lung volumes. Other activities examine factors such as airway resistance and the effect of surfactant on lung function.

Exercise 8 Chemical and Physical Processes of Digestion. Turns the student's computer into a virtual chemistry lab where enzymes, reagents, and incubation conditions can be manipulated (in compressed time) to examine factors that affect enzyme activity.

Exercise 9 Renal System Physiology. Simulates the function of a single nephron. Topics include factors influencing glomerular filtration, the effect of hormones on urine function, and glucose transport maximum.

Exercise 10 Acid-Base Balance. Topics include respiratory and metabolic acidosis/alkalosis, as well as renal and respiratory compensation.

Using the Histology Module. Includes over 200 histology images, viewable at various magnifications, with accompanying descriptions and labels. A new Histology Review Supplement has also been added, consisting of forty slides that relate to topics covered in the PhysioEx lab simulations.

Getting Started

To use PhysioEx version 4.0, your computer should meet the following minimum requirements.

- **IBM/PC:** Windows 95, 98, NT 2000, Millennium Edition or higher; Pentium I/266 mHz or faster.

- **Macintosh:** Macintosh 8.6 and above; 604/300 mHz or G3/233 mHz.

- 64 MB RAM (128 MB recommended)

- 800 × 600 screen resolution, millions of colors

- Internet Explorer 5.0 (or higher) *or* Netscape 4.6 (or higher)*

- Flash 6† plug-in

- 4× CD-ROM drive (if using CD-ROM version)

- Printer

*CD version users: Although you do not need a live Internet connection to run the CD, you do need to have a browser installed on your computer. If you do not have a browser, the CD includes a free copy of Netscape which you may install. See instructions on the CD-liner notes.

†CD version users: If you do not have Flash 6 installed on your computer, the CD includes a free Flash installer. See instructions on the CD-liner notes.

Instructions for Getting Started—Mac Users (CD version)

1. Put the PhysioEx CD in the CD-ROM drive. The program should launch automatically. If autorun is disabled on your computer, double click on the PhysioEx icon that appears on your desktop.

2. Although you do not need a live Internet connection to run PhysioEx, you do need to have a browser (such as Netscape or Internet Explorer) installed on your computer. If you already have a browser installed, proceed to step 3. If you do not have a browser installed, follow the instructions for installing Netscape found on the liner notes that are packaged with your CD.

3. If you can see a clock onscreen with the clock hands moving, click Proceed. If you cannot see the clock, or if you can see the clock but the hands are not moving, follow the instructions for installing Flash 6 found on the liner notes that are packaged with your CD.

4. On the License Agreement screen, click **Agree** to proceed.

5. On the screen with the PhysioEx icon at the top, click the **License Agreement** link to read the full agreement. Then close the License Agreement window and click the **Main Menu** link.

6. From the Main Menu, click on the lab you wish to enter.

Instructions for Getting Started—IBM/PC Users (CD version)

1. Put the PhysioEx CD in the CD-ROM drive. The program should launch automatically. If autorun is disabled on your machine, double click the My Computer icon on your Windows desktop, and then double click the PhysioEx icon.

2. Although you do not need a live Internet connection to run PhysioEx, you do need to have a browser (such as Netscape or Internet Explorer) installed on your computer. If you already have a browser installed, proceed to step 3. If you do not have a browser installed, follow the instructions for installing Netscape found on the liner notes that are packaged with your CD.

3. If you can see a clock onscreen with the clock hands moving, click **Proceed.** If you cannot see the clock, or if you can see the clock but the hands are not moving, follow the instructions for installing Flash 6 found on the liner notes that are packaged with your CD.

4. On the License Agreement screen, click **Agree** to proceed.

5. On the screen with the PhysioEx icon at the top, click the **License Agreement** link to read the full agreement. Then close the License Agreement window and click the **Main Menu** link.

6. From the Main Menu, click on the lab you wish to enter.

Instructions for Getting Started—Web Users

If you have the web version of this lab manual, follow the instructions for accessing www.physioex.com that appear at the very front of this booklet.

Technical Support

Phone: 800.677.6337
Email: media.support@pearsoned.com
Hours: 8 A.M. to 5 P.M. CST, Monday–Friday

Cell Transport Mechanisms and Permeability

Objectives

1. To understand the selective permeability function of the plasma membrane
2. To be able to describe the various mechanisms by which molecules may passively cross the plasma membrane
3. To be able to describe the various mechanisms by which molecules are actively transported across the plasma membrane
4. To understand the differences between how membrane transporters work with and without the expenditure of cellular metabolic energy
5. To define **passive transport, active transport, simple diffusion, facilitated diffusion, osmosis, solute pump, hypotonic, isotonic,** and **hypertonic**

Each cell in your body is surrounded by a plasma membrane that separates the cell from interstitial fluid. The major function of the plasma membrane is to selectively permit the exchange of molecules between the cell and the interstitial fluid, so that the cell is able to take in substances it needs while expelling the ones it does not. These substances include gases, such as oxygen and carbon dioxide; ions; and larger molecules such as glucose, amino acids, fatty acids, and vitamins.

Molecules move across the plasma membrane either *passively* or *actively*. In **active transport,** molecules move across the plasma membrane with the expenditure of cellular energy (ATP). In **passive transport,** molecules pass through the plasma membrane without the expenditure of any cellular energy. Examples of passive transport are **simple diffusion, osmosis,** and **facilitated diffusion. Simple diffusion** is the spontaneous movement of molecules across a biological membrane's lipid bilayer from an area of higher concentration to an area of lower concentration. **Osmosis** is the diffusion of water across a semipermeable membrane. **Facilitated diffusion** is the movement of molecules across a selectively permeable membrane with the aid of specialized transport proteins embedded within the membrane.

In this lab, we will be simulating each of these cell transport mechanisms. We will begin by examining simple diffusion.

Simple Diffusion

All molecules, whether solid, liquid, or gas, are in continuous motion or vibration. If there is an increase in temperature, the molecules will move faster. The moving molecules bump into each other, causing each other to alter direction. Thus, the movement of molecules is said to be "random." If one were to release a drop of liquid food coloring into a large beaker of water, the food coloring molecules would randomly move until their concentration was equal throughout the beaker. The molecules would reach equilibrium through this process of *diffusion*. We define diffusion as the movement of molecules from one location to another as a result of their random thermal motion. **Simple diffusion** is diffusion across a biological membrane's lipid bilayer.

The speed at which a molecule moves across a membrane depends in part on the mass, or molecular weight, of the molecule. The higher the mass, the slower the molecule will diffuse. Normally, the rate at which a substance diffuses across the membrane can be determined by measuring the rate at which the concentration of the substance on one side of the membrane approaches the concentration of the substance on the other side of the membrane. The magnitude of the net movement across the membrane, or flux (F), is proportional to the concentration difference between the two sides of the membrane ($C_o - C_i$), the surface area of the membrane (A), and the membrane permeability constant (k_p):

$$F = k_pA(C_o - C_i)$$

Nonpolar substances will diffuse across a membrane fairly rapidly. The reason is that nonpolar substances will dissolve in the nonpolar regions of the membrane—regions that are occupied by the fatty acid chains of the membrane phospholipids. Gases such as oxygen and carbon dioxide, steroids, and fatty acids are prime nonpolar molecules that will diffuse through a membrane rapidly.

In contrast, polar substances have a much lower solubility in the membrane phospholipids. Certain compounds that are intermediates of metabolism are not usually allowed through the membrane, as they are often ionized and contain groups such as phosphate. Thus, once produced in a cell, they cannot leave even if their concentrations are higher inside the cell than they are outside the cell. From this we can see that it is the lipid bilayer portion of the plasma membrane that is responsible for the membrane's selectivity in what it allows through.

Ions, such as Na^+ and Cl^-, tend to diffuse across a membrane rather rapidly. This suggests that a protein component of the membrane is involved—and in fact, proteins do form channels that allow these ions to pass from one side of the membrane to the other. Remember that the channels are selective. Channels that allow sodium through will not usually allow other ions, such as calcium, through.

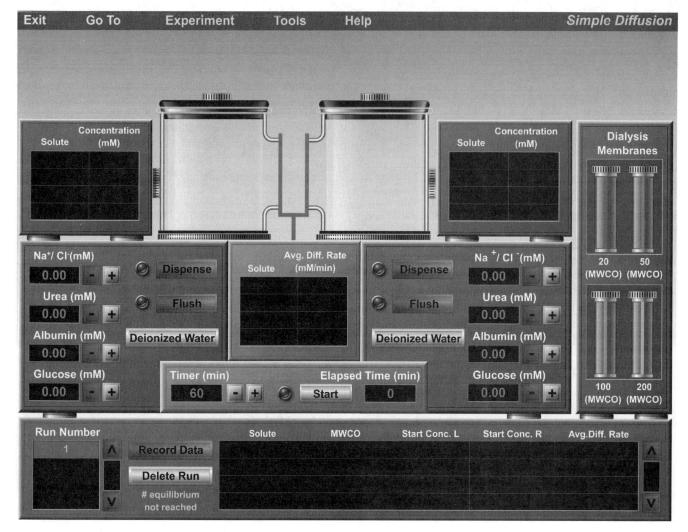

Figure 1.1 Opening screen of the Simple Diffusion experiment.

Diffusion will lead to a state in which the concentration of the diffusing solutes is constant in space and time. Diffusion across a membrane tends to equilibrate so that there are equal solute concentrations on both sides of the membrane. The rate of diffusion is proportional to both the area of the membrane and the difference in concentration of the solute on both sides of the membrane. **Fick's first law of diffusion** states

$$J = -DA \, \Delta_c/\Delta_x$$

where

J = net rate of diffusion (gms or mols/unit time)
D = diffusion coefficient for the diffusing solute
A = area of the membrane
Δ_c = concentration difference across the membrane
Δ_x = thickness of the membrane

Activity 1:
Simulating Simple Diffusion

Follow the instructions in the "Getting Started" section at the front of this manual for starting PhysioEx 3.0. From the Main Menu, select the first lab: **Cell Transport Mechanisms and Permeability.** You will see the opening screen for the "Simple Diffusion" activity, shown in Figure 1.1.

In this activity we will be simulating the process of diffusion across the plasma membrane. Notice the two glass beakers at the top of the screen. You will be filling each beaker with fluid. Imagine that the right beaker represents the inside of a cell, while the left beaker represents the extracellular (interstitial) fluid. Between the two beakers is a membrane holder into which you will place one of four dialysis membranes found on the right side of the screen. Each of these membranes has a different "MWCO," which stands for "molecular weight cut off." Molecules with a molecular weight of less than this value may pass through the membrane, while molecules with higher molecular weight values cannot. To move a membrane to the membrane holder, click on the membrane, drag it to the membrane holder, and let go of your mouse button—the membrane will lock into place between the two beakers.

Below each of the two beakers is a solutions dispenser. You may set how many millimoles (mM) of different solutes (Na^+/Cl^-, urea, albumin, or glucose) you want to dispense into each beaker by clicking on the (+) or (−) buttons beneath each solute name. You may also dispense deionized water into either beaker by clicking the **Deionized Water** button under the beaker you wish to fill. Clicking the **Dispense** buttons under each beaker will then cause the beakers to fill with fluid. Clicking the **Flush** buttons under each beaker will empty the beakers.

At the bottom of the screen is a data recording box. After each experimental run, you may record your data by clicking the **Record Data** button. If you wish to delete the data for any given run, simply highlight the line of data you wish to delete and then click **Delete Run.** You may also print out your data by clicking Tools (at the top of the screen) and then selecting **Print Data.**

1. Using the mouse, click on the dialysis membrane with the MWCO of 20 and drag it into the membrane holder.

2. Adjust the mM concentration of Na^+/Cl^- for the left beaker to 9 mM by clicking the (+) button. Then click the **Dispense** button under the left beaker to fill the beaker.

3. Click the **Deionized Water** button under the right beaker and click **Dispense** under the right beaker to fill the beaker.

4. Set the Timer for 60 minutes by clicking the (+) button next to the Timer display (which will be compressed into 60 seconds.)

5. Click on the **Start** button to start the experimental run. Note that the membrane container descends into the equipment. Also note that the **Start** button is now a **Pause** button, which you may click to pause any run.

6. As the Elapsed Time display reaches 60, note the concentration readings for each beaker in the displays on each side of the two beakers.

7. Once the Elapsed Time display has reached 60, you will see a dialogue box pop up telling you whether or not equilibrium was reached.

8. Click **Record Data** to save the data from this run.

9. Click the **Flush** buttons on both the left and right sides to empty the beakers.

10. Return the dialysis membrane to its starting place by clicking and dragging it back to the membrane chamber.

11. Now, repeat steps 1–10 with each of the remaining dialysis membranes. Be sure to record the data for each run. After each run, flush both vessels and return the dialysis membrane.

Turn to the Periodic Table of Elements on p. 4 of this booklet.

What is the molecular weight of Na^+? _____

What is the molecular weight of Cl^-? _____

Which MWCO dialysis membranes allowed both of these ions through? _____

12. Repeat this experiment using each of the remaining solutes (urea, albumin, and glucose) in the left beaker and deionized water in the right beaker. Be sure to click **Record Data,** flush both beakers, and replace the dialysis membrane after each run. Click **Tools → Print Data** to print your data.

13. Fill in the chart below with your results.

Chart 1 Did Diffusion Take Place?				
Solute	**Membrane (MWCO)**			
	20	50	100	200
NaCl				
Urea				
Albumin				
Glucose				

Which materials diffused from the left beaker to the right beaker?

_____

Which did not?

_____

Why?

_____

_____

_____

_____

Periodic Table of Elements

Key:

1	Atomic number
H	Symbol
Hydrogen	Element
1.00794	Atomic weight

IA 1	IIA 2	IIIB 3	IVB 4	VB 5	VIB 6	VIIB 7	VIIIB 8	VIIIB 9	VIIIB 10	IB 11	IIB 12	IIIA 13	IVA 14	VA 15	VIA 16	VIIA 17	VIIIA 18
1 **H** Hydrogen 1.008																	**2** **He** Helium 4.003
3 **Li** Lithium 6.941	**4** **Be** Beryllium 9.012											**5** **B** Boron 10.81	**6** **C** Carbon 12.01	**7** **N** Nitrogen 14.01	**8** **O** Oxygen 16.00	**9** **F** Fluorine 19.00	**10** **Ne** Neon 20.18
11 **Na** Sodium 22.99	**12** **Mg** Magnesium 24.31											**13** **Al** Aluminum 26.98	**14** **Si** Silicon 28.09	**15** **P** Phosphorus 30.97	**16** **S** Sulfur 32.06	**17** **Cl** Chlorine 35.45	**18** **Ar** Argon 39.95
19 **K** Potassium 39.10	**20** **Ca** Calcium 40.08	**21** **Sc** Scandium 44.96	**22** **Ti** Titanium 47.88	**23** **V** Vanadium 50.94	**24** **Cr** Chromium 52.00	**25** **Mn** Manganese 54.94	**26** **Fe** Iron 55.85	**27** **Co** Cobalt 58.93	**28** **Ni** Nickel 58.70	**29** **Cu** Copper 63.55	**30** **Zn** Zinc 65.38	**31** **Ga** Gallium 69.72	**32** **Ge** Germanium 72.59	**33** **As** Arsenic 74.92	**34** **Se** Selenium 78.96	**35** **Br** Bromine 79.90	**36** **Kr** Krypton 83.80
37 **Rb** Rubidium 85.47	**38** **Sr** Strontium 87.62	**39** **Y** Yttrium 88.91	**40** **Zr** Zirconium 91.22	**41** **Nb** Niobium 92.91	**42** **Mo** Molybdenum 95.94	**43** **Tc** Technetium (98)	**44** **Ru** Ruthenium 101.1	**45** **Rh** Rhodium 102.9	**46** **Pd** Palladium 106.4	**47** **Ag** Silver 107.9	**48** **Cd** Cadmium 112.4	**49** **In** Indium 114.8	**50** **Sn** Tin 118.7	**51** **Sb** Antimony 121.8	**52** **Te** Tellurium 127.6	**53** **I** Iodine 126.9	**54** **Xe** Xenon 131.3
55 **Cs** Cesium 132.9	**56** **Ba** Barium 137.3	**57** **La*** Lanthanum 138.9	**72** **Hf** Hafnium 178.5	**73** **Ta** Tantalum 180.9	**74** **W** Tungsten 183.9	**75** **Re** Rhenium 186.2	**76** **Os** Osmium 190.2	**77** **Ir** Iridium 192.2	**78** **Pt** Platinum 195.1	**79** **Au** Gold 197.0	**80** **Hg** Mercury 200.6	**81** **Tl** Thallium 204.4	**82** **Pb** Lead 207.2	**83** **Bi** Bismuth 209.0	**84** **Po** Polonium (209)	**85** **At** Astatine (210)	**86** **Rn** Radon (222)
87 **Fr** Francium (223)	**88** **Ra** Radium (226.0)	**89** **Ac**** Actinium (227)	**104** **Rf** Rutherfordium (261)	**105** **Ha** Hahnium (262)	**106** **Unh** Unnilhexium (263)	**107** **Uns** Unnilseptium (262)	**108** **Uno** Unniloctium (265)	**109** **Une** Unnilennium (266)									

*Lanthanides

58 **Ce** Cerium 140.1	**59** **Pr** Praseodymium 140.9	**60** **Nd** Neodymium 144.2	**61** **Pm** Promethium (145)	**62** **Sm** Samarium 150.4	**63** **Eu** Europium 152.0	**64** **Gd** Gadolinium 157.3	**65** **Tb** Terbium 158.9	**66** **Dy** Dysprosium 162.5	**67** **Ho** Holmium 164.9	**68** **Er** Erbium 167.3	**69** **Tm** Thulium 168.9	**70** **Yb** Ytterbium 173.0	**71** **Lu** Lutetium 175.0

**Actinides

90 **Th** Thorium 232.0	**91** **Pa** Protactinium (231)	**92** **U** Uranium 238.0	**93** **Np** Neptunium (237)	**94** **Pu** Plutonium (244)	**95** **Am** Americium (243)	**96** **Cm** Curium (247)	**97** **Bk** Berkelium (247)	**98** **Cf** Californium (251)	**99** **Es** Einsteinium (252)	**100** **Fm** Fermium (257)	**101** **Md** Mendelevium (258)	**102** **No** Nobelium (259)	**103** **Lr** Lawrencium (260)

Activity 2:
Simulating Dialysis

Now, let's set up a mock dialysis machine experiment. These machines are used on patients who have lost kidney function. Urea, a breakdown product of amino acids, must be removed from the patient's blood or it will become toxic to the body and cause death. Dialysis machines take a patient's blood and pass it through a selectively permeable membrane in order to remove urea from the blood. On one side of the membrane is the patient's blood; on the other side are fluids carefully selected to mimic the concentrations found in the body of substances such as Na^+, K^+, Ca^{++} and HCO_3^-. To simulate this process:

1. Place the dialysis membrane of 200 MWCO into the membrane holder.

2. Set up the left beaker with 10 mM of each of the four solutes and dispense. This beaker will represent the dialysis patient's blood.

3. Set up the right beaker the same way, except set the urea concentration at 0 mM—in other words, the right beaker will contain no urea.

4. Set the Timer for 60 minutes, then click **Start** and wait for the experimental run to complete.

What happens to the urea concentration in the left beaker (the patient)?

_____

_____

Why does this occur?

_____

_____

_____

Normally, dialysis machines are set to run so that the blood is subjected to diffusion twice, and urea is reduced by 75% rather than 50%. In addition, excess water is drawn from the patient, who has no other way to dispose of excess fluid. Dialysis patients need to have routine lab tests done to ensure that ion concentrations are maintained at normal levels.

Facilitated Diffusion

Simple diffusion accounts for the transmembrane transport of some ions, but not all of them. Some molecules that are too polar to diffuse still manage to get through the plasma membrane's lipid bilayer. Similarly, some molecules that are too large to pass through protein channels still manage to cross the membrane. How? The passage of such molecules and the nondiffusional movement of ions through a membrane is mediated by integral proteins known as **transporters.** Transporters are embedded within the plasma membrane and work by undergoing a conformational change that allows transport to occur. A molecule first binds to a receptor site on a transporter. When bound, the transporter changes shape so that the binding site moves from one side of the membrane to the other side. The molecule then dissociates from the transporter and is released on the other side of the membrane. This type of transport is called **facilitated diffusion.** It is considered a form of passive transport because no cellular energy is expended in the process.

The term _facilitated diffusion_ is a bit misleading since the process does not really involve diffusion (which, you will recall, is the movement of molecules from one location to another along a concentration gradient, as a result of random thermal motion). In facilitated diffusion, molecules are still moving from one location to another along a concentration gradient, but it is transport proteins that result in this movement—not random thermal motion. The end results of diffusion and facilitated diffusion are the same. The net flux proceeds from an area of high concentration to an area of low concentration until the concentrations are equal on both sides of the membrane.

Among the most important facilitated-diffusion systems in the body are those that move glucose across the membrane. Without transporters, the relatively large, polar glucose molecule would never be able to pass into a cell. However, the number of transport proteins in a given cell membrane is finite, so only a certain amount of glucose can be transported per unit of time. Transport of glucose into the cell is especially interesting in that the glucose is converted to glucose-6-phosphate as soon as it enters the cell, so that there is always a low concentration of glucose inside the cell, which favors transport into the cell. ■

Activity 3:
Facilitated Diffusion

Using the mouse, click on **Experiment** at the top of the screen. A drop-down menu will appear. Select **Facilitated Diffusion.** A new screen will appear (see Figure 1.2). You will notice two key changes from the first screen. First, in place of the dialysis membranes on the right side of the screen, there is now a "membrane builder." This will be used to "make" membranes that will transport molecules from one beaker to the other. The second change is that in this experiment, we will be working with glucose and Na^+/Cl^- solutes only.

1. Note that the **Glucose Carriers** display is currently set at 500. Click on **Build Membrane** in order to create a membrane with 500 glucose carriers.

2. Click and drag this membrane to the membrane holder between the two beakers.

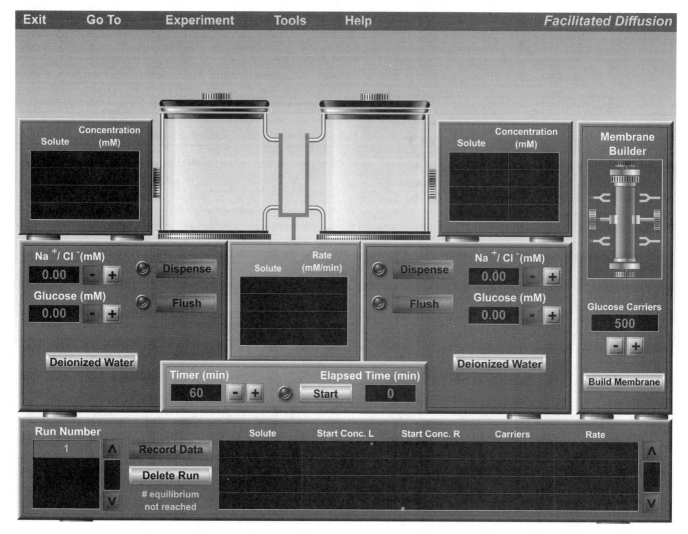

Figure 1.2 Opening screen of the Facilitated Diffusion experiment.

3. For the left beaker, set Na$^+$/Cl$^-$ to 9 mM and glucose to 9 mM by clicking on the corresponding (+) buttons. Then click **Dispense** to fill up the left beaker.

4. For the right beaker, click on the **Deionized Water** button below the beaker and then click **Dispense.**

5. Set the timer for 60 minutes and click **Start.**

6. Allow the run to complete. When the Elapsed Timer reaches 60, click on **Record Data** to record your data. Also record your data in Chart 2 .

7. Click the **Flush** button under each beaker to empty the beakers, and return the membrane to the membrane builder.

8. Build a new membrane with 300 glucose carriers and repeat this experiment. Be sure to record your results, flush the beakers, and replace the membrane after each run.

9. Build a membrane with 700 glucose carriers and repeat the experiment.

10. Build a membrane with 900 glucose carriers and repeat the experiment.

11. For comparison, lower the glucose concentration to 3 mM and repeat steps 1–10 of the experiment. Record your results after each run by clicking **Record Data** and by filling in Chart 2 below.

Chart 2   Facilitated Diffusion Results

Glucose Concentration (mM)	No. of glucose carrier proteins			
	300	500	700	900
3				
9				

12. Click **Tools → Print Data** to print your data.

At a given glucose concentration, how does the amount of time it takes to reach equilibrium change with the number of carriers used to "build" the membrane?

_____

Does the diffusion rate of Na^+/Cl^- change with the number of receptors?

_____

What is the mechanism of the Na^+/Cl^- transport?

_____

If you put the same amount of glucose in the right beaker as in the left, would you be able to observe any diffusion?

_____

Does being unable to observe diffusion necessarily mean that diffusion is not taking place?

_____

_____

_____ ■

Osmosis

A *semipermeable membrane* is a membrane that is permeable to water but not to solutes. **Osmosis** is defined as the flow of water across a semipermeable membrane from an area of higher water concentration (lower solute concentration) to an area of lower water concentration (higher solute concentration). The greater the solute concentration, the lower the water concentration. Osmosis is further defined as a "colligative property" because it depends on solute concentration rather than solute chemical properties. Water is a small, polar molecule that diffuses across cell membranes very rapidly. Because of its polar nature, one might expect that water would not penetrate the nonpolar lipid regions of the cell membrane. Membrane proteins, called *aquaporins,* form channels through which water can diffuse. The concentration of these aquaporins varies with tissue type.

It is essential to understand that the degree to which water concentration is decreased by addition of solute depends on the number of solute particles added. For example, 1 mol of glucose decreases the water concentration approximately the same as a 1 mol solution of amino acid or 1 mol of urea. A molecule that ionizes decreases the water concentration in proportion to the number of ions formed. Therefore, a 1 mol solution of Na^+/Cl^- produces a 1 mol solution of Na^+ plus a 1 mol solution of Cl^-. Therefore, it is basically a 2 mol solution.

Two beakers separated by a dialysis membrane (such as the ones we have been working with) are not infinitely expandable. The transfer of water from one compartment to the other will increase the amount of water in the second compartment. If the limits of the beaker cannot expand, pressure within the second beaker will increase, eventually preventing further water entry. The amount of pressure that needs to be supplied to the second beaker in order to prevent further water entry from the first beaker is called *osmotic pressure.* Osmotic pressure is another characteristic that depends on the solution's water concentration.

If the solutions in the beakers have the same concentration of nonpenetrating solutes on either side of the membrane, the two solutions are said to be **isotonic** (iso = same). Solutes that are "penetrating" do not contribute to the tonicity of a solution as they pass from one side of the membrane to the next with no problems. When two solutions are compared and one has a lower concentration of solutes, that solution is said to be **hypotonic** (hypo = less). The other solution, the one with the higher concentration, is said to be **hypertonic** (hyper = more). This is important when discussing cells. If a cell is hypertonic to its surrounding medium, water will flow into the cell to dilute the hypertonic solution. Often, so much water enters the cell that the cell bursts.

Activity 4:
Osmosis

Click on **Experiment** at the top of the screen and then select **Osmosis.** A new screen will appear (Figure 1.3). The screen is similar to the one we saw for the **Simple Diffusion** experiment. The main change is that on top of each beaker is a pressure indicator, which we will be watching during experimental runs.

1.   Drag the 20 MWCO membrane and place it between the two beakers.

2.   Set the Na^+/Cl^- concentration for the left beaker at 9 mM and click **Dispense.**

3.   Fill the right beaker with **Deionized Water** and click Dispense.

4.   Set the Timer for 60 minutes.

5.   Click on **Start** and allow the experiment to run. Pay attention to the "Pressure" indicators on top of each beaker.

6.   Once the Elapsed Time is up, click Record Data. Record the data in Chart 3 on p. 8 as well.

7.   Click **Flush** under both beakers to empty them.

8.   Return the membrane to its original place.

9.   Repeat the experiment using the remaining three membranes. Be sure to record all of your data, flushing the beakers in between each run.

Did you observe any pressure changes during this experiment? If so, in which beaker(s), and with which membranes?

_____

_____

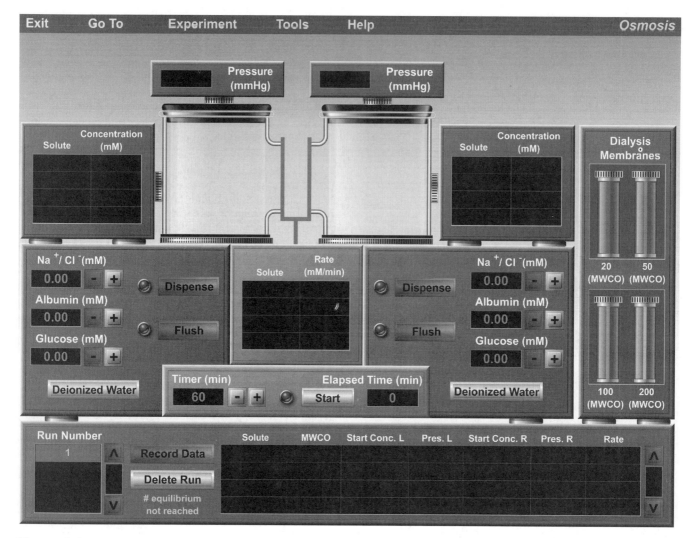

Figure 1.3   Opening screen of the Osmosis experiment.

Why?

_____

_____

_____

Did the Na$^+$/Cl$^-$ diffuse from the left beaker to the right beaker? If so, with which membrane(s)?

_____

_____

Why?

_____

_____

_____

10. Repeat the experiment, first using 9 mM albumin in the left beaker, then 9mM glucose. Click **Record Data** after each run; also record your data in Chart 3.

Chart 3 Osmosis Results (pressure in mm Hg)

Solute	Membrane (MWCO)			
	20	50	100	200
Na$^+$/Cl$^-$				
Albumin				
Glucose				

11. Click **Tools → Print Data** to print your data.

Explain the relationship between solute concentration and osmotic pressure.

_____

_____

Does diffusion allow osmotic pressure to be generated?

_____

_____

Would pressure be generated if solute concentrations were equal on both sides of the membrane?

_____

Why or why not?

_____

_____

Would pressure be generated if you had 9 mM glucose on one side of a 200 MWCO membrane and 9 mM NaCl on the other side? If so, which solution was generating the pressure?

_____

_____

Would pressure be generated if you had 9 mM albumin on one side of a 200 MWCO membrane and 9 mM NaCl on the other side? If so, which solution was generating the pressure?

_____

_____ ■

Filtration

At the same time that diffusion is allowing cells to take in oxygen and nutrients while expelling carbon dioxide and metabolic wastes, another process is also taking place. This process occurs mainly in capillaries of the body (such as those in the kidneys) where fluid pressure of the blood—called _hydrostatic_ pressure—forces materials across a capillary wall. Both blood and interstitial fluid contain dissolved solutes. Usually, the osmotic pressure of the interstitial fluid is not as great as the hydrostatic pressure of the blood, so there is a net movement of fluid and/or solutes out of capillaries—a process called **filtration.** What is filtered out depends solely on the molecular size of the solute and the size of the "pores" in the membrane. Filtration is considered a passive process, since it occurs without the expenditure of metabolic energy.

A c t i v i t y   5 :
Filtration

Click on **Experiment** at the top of the screen and select **Filtration.** You will see an opening screen that looks noticeably different from the earlier activities (Figure 1.4). Note the two beakers situated on the left side of the screen, one on top of the other. Note also that the top beaker contains a pressure gauge. Unlike the Osmosis experiment, in which the pressure gauge detected pressure developed due to water movement, this pressure gauge measures the hydrostatic pressure that will filter fluid from the top beaker into the bottom beaker. Finally, note the "Membrane Residue Analysis" box. This will be used to detect if any solutes are left on a membrane after each experimental run.

1. Click and drag the 20 MWCO membrane into the membrane holder between the two beakers.

2. Set Na^+/Cl^- to 9 mM, urea and glucose to 5 mM, and powdered charcoal to 5 mg/ml by clicking the (+) button next to each solute. Then click **Dispense** to dispense into the upper beaker.

3. Leave the pressure at 50 mm Hg and the timer at 60 minutes, the default settings. Click on **Start.** You will see fluid being filtered into the bottom beaker.

4. Watch the Filtrate Analysis Unit (next to the bottom beaker) for any activity. This will tell you which solutes are passing through the membrane.

5. When the 60 minutes are up, drag the membrane to the Membrane Residue Analysis unit and let go of your mouse. The membrane will lock into place. Click on **Start Analysis.** In the data box below, you will see what solute(s) were detected on the membrane used for filtration.

6. Record your data by clicking **Record Data.**

What were the results of your initial membrane analysis?

_____

7. Click **Flush** and return the membrane to its original location.

8. Drag the 50 MWCO membrane to the membrane holder between the beakers.

9. Leave the pressure at 50 and repeat the experiment. When the timer has reached 60 minutes, perform a membrane analysis and click **Record Data.**

10. Click **Flush** and return the membrane

11. Repeat steps 8–10 with the remaining two membranes. Be sure to record your data for each run.

12. Increase the pressure to 100 mm Hg and repeat the entire experiment. Again, record all experimental data.

13. Click **Tools → Print Data** to print your data.

Does the membrane MWCO affect filtration rate?

_____

Does the amount of pressure applied affect the filtration rate?

_____

Did all solutes pass through all the membranes?

_____

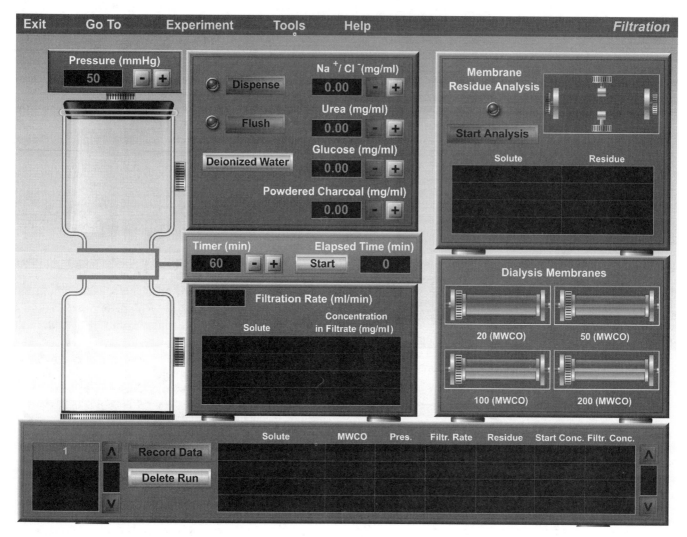

Figure 1.4 Opening screen of the Filtration experiment.

If not, which one(s) did not?

_____

Why?

_____

How can the body selectively increase the filtration rate of a given organ or organ system?

_____

_____

_____

Active Transport

Active transport differs from passive transport in that energy derived from metabolism is used to move solutes across the membrane. It also differs in that solutes are moved from an area of low concentration to an area of high concentration—the opposite of facilitated diffusion. As with facilitated diffusion, binding of a substance to a transporter is required. Since the bound substance is moving "uphill" to an area of higher concentration, the transporters are often spoken of as **pumps.** The net movement from lower to higher concentration and the maintenance of a higher steady-state concentration on one side of a membrane can be achieved only by the continuous input of energy into the active-transport mechanism. The energy input can alter the affinity of the binding site on the transporter so that there is a higher affinity when facing one direction over the other, or the energy may alter the rates at which the transporter moves the binding site from one side of a membrane to the other. As with facilitated diffusion, the number of transport molecules per cell is finite.

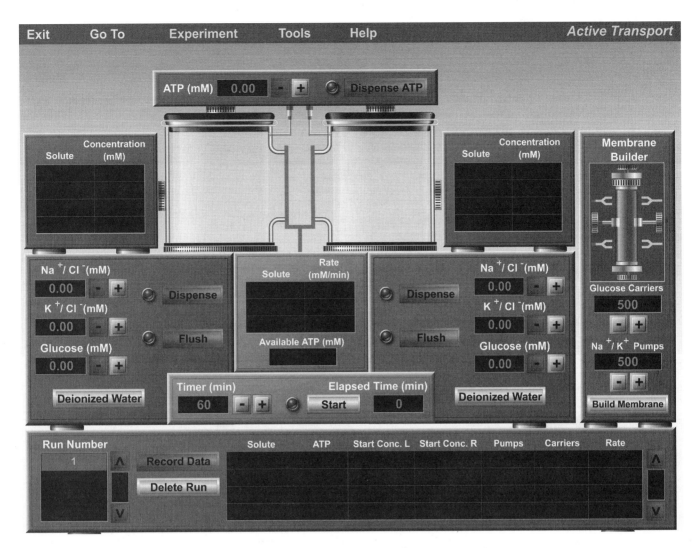

Figure 1.5 Opening screen of the Active Transport experiment.

Energy for active transport is derived from cellular metabolism. Inhibition of ATP blocks the active-transport mechanism. In order for solutes to be moved from an area of lower concentration to an area of higher concentration, the transport must be coupled with the flow of energy from a higher energy level to a lower energy level. If ATP is used directly in the transport, the transport mechanism is known as *primary active transport.*

Energy is derived from hydrolysis of ATP by a transporter which is an ATPase that catalyzes the breakdown of ATP and phosphorylates itself. This phosphorylation of the transporter will either alter the affinity of the binding site or the rate of conformational change. Four primary active transport proteins have been identified. In all plasma membranes, there is the sodium-potassium ATPase, responsible for the outward flow of sodium and inward flow of potassium. Sodium is the primary ion found in the extracellular fluid, while potassium is the ion found, for the most part, inside cells. Other transport proteins are involved with calcium transport, hydrogen transport, and hydrogen-potassium transport.

Activity 6:
Active Transport

Click on **Experiment** at the top of the screen and select **Active Transport.** A new screen will appear that resembles the screen from facilitated diffusion (Figure 1.5). The key change is the addition of an ATP dispenser on top of the beakers. Remember, since ATP is needed for the system to run, it must be added for each run.

1. In the membrane builder, be sure that the number of glucose carriers is set at 500 and that the number of Na^+/Cl^- pumps is also set at 500.

2. Click on **Build Membrane.**

3. Drag the "built" membrane to the membrane holder between the two beakers.

4. For the left beaker, set Na^+/Cl^- to 9 mM by clicking the (+) button and click **Dispense.**

5. For the right beaker, click **Deionized Water** and then click **Dispense.**

6. Set ATP to 1 mM and then click **Dispense ATP.**

7. Be sure the Timer is set at 60 minutes, and then click **Start.**

At the end of this experimental run, did the Na^+/Cl^- move from the left vessel to the right vessel?

_____

_____

Why?

_____

8. Click **Flush** under both beakers.

9. Add 9 mM Na^+/Cl^- to the left beaker and 9 mM KCl to the right beaker.

10. Set ATP to 1 mM, click **Dispense ATP** and click **Start.**

11. At the end of the run, click **Record Data.**

As the run progresses, the concentrations of the solutes will change in the windows next to the two beakers. The rate will slow down markedly, then stop before completed. Why?

_____

_____

Now that you have performed the basic experiment, let's conduct two variations.

12. Repeat the experiment, except increase the amount of ATP added to the system.

Does the amount of NaCl/KCl transported change?

_____

13. Repeat the experiment, except change the number of carriers and pumps when you build the membrane.

Does the amount of solute transported across the membrane change with an increase in carriers or pumps?

_____

Is one solute more affected than the other?

_____

Does the membrane you "built" allow simple diffusion?

_____

If you placed 9 mM NaCl on one side of the membrane and 15 mM on the other side, would there be movement of the NaCl?

_____

_____

Why?

_____

Does the amount of ATP added make any difference?

_____

_____

14. Click **Tools → Print Data** to print your recorded data. ■

Skeletal Muscle Physiology

Objectives

1. To define **motor unit, twitch, latent period, contraction phase, relaxation phase, threshold, summation, tetanus, fatigue, isometric contraction,** and **isotonic contraction**
2. To understand how nerve impulses trigger muscle movement
3. To describe the phases of a muscle twitch
4. To identify threshold and maximal stimuli
5. To understand the effect of increases in stimulus intensity on a muscle
6. To understand the effect of increases in stimulus frequency on a muscle
7. To demonstrate muscle fatigue
8. To explain the differences between isometric and isotonic muscle contractions

Humans make voluntary decisions to talk, walk, stand up, or sit down. The muscles that make these actions possible are skeletal muscles. Skeletal muscle is generally muscle that is attached to the skeleton of the body, although there are some exceptions—for example, the *obicularis oris,* a muscle around the mouth, never attaches to any skeletal element. Skeletal muscles characteristically span two joints and attach to the skeleton via tendons connecting to the periosteum of the bone.

The Motor Unit and Muscle Contraction

A **motor unit** consists of a motor neuron and all of the muscle fibers it innervates. Motor neurons direct muscles when and when not to contract. A motor neuron and a muscle cell intersect at what is called the *neuromuscular junction.* Specifically, the neuromuscular junction is where the axon terminal of the neuron meets a specialized region of the muscle cell's plasma membrane. This specialized region is called the *motor endplate.* An action potential (depolarization) in a motor neuron triggers the release of acetylcholine, which diffuses into the muscle plasma membrane (also known as the *sarcolemma*). The acetylcholine binds to receptors on the muscle cell, initiating a change in ion permeability that results in depolarization of the muscle plasma membrane, called an *end-plate potential.* The end-plate potential, in turn, triggers a series of events that results in the contraction of a muscle cell. This entire process is called *excitation-contraction coupling.*

We will be simulating this process in the following activities, only instead of using acetylcholine to trigger action potentials, we will be using electrical shocks. The shocks will be administered by an electrical stimulator that can be set for the precise voltage, frequency, and duration of shock desired. When applied to a muscle that has been surgically removed from an animal, a single electrical stimulus will result in a muscle **twitch**—the mechanical response to a single action potential. A twitch has three phases: the **latent period,** which is the period of time that elapses between the generation of an action potential in a muscle cell and the start of muscle contraction; the **contraction phase,** which starts at the end of the latent period and ends when muscle **tension** peaks; and the **relaxation phase,** which is the period of time from peak tension until the end of the muscle contraction.

Single Stimulus

Follow the instructions in the "Getting Started" section at the front of this manual for starting PhysioEx 3.0. From the Main Menu, select **Skeletal Muscle Physiology.** You will see the opening screen for the **Single Stimulus** activity, shown in Figure 2.1.

On the left side of the screen is a muscle suspended in a metal holder that is designed to measure any force produced by the muscle. To the right of the metal holder are three pieces of equipment. The top piece of equipment is an oscilloscope screen. When you apply an electrical stimulus to the muscle, the muscle's reaction will be graphically displayed on this screen. Elapsed time, in milliseconds, is measured along the X axis of this screen, while any force generated by the muscle is measured along the Y axis. In the lower right-hand corner of the oscilloscope is a **Clear Tracings** button; clicking the button will remove any tracings from the screen.

Beneath the oscilloscope screen is the electrical stimulator you will use to stimulate the muscle. Note the electrode from the stimulator that rests on the muscle. Next to the **Voltage** display on the left side of the stimulator are (+) and (−) buttons, which you may click to set the desired voltage. When you click on the **Stimulate** button, you will electrically stimulate the muscle at the set voltage. In the middle of the stimulator are display fields for *active force, passive force,* and *total force.* Muscle contraction produces *active force. Passive force* is generated from the muscle being stretched. The sum of active force and passive force is the *total force.* Also notice a **Measure** button on the stimulator. Clicking this button after administering a stimulus will cause a yellow vertical line to appear. Clicking the (+) or (−) buttons under **Time (msec)** will then allow you to move the yellow line along the X axis and view the active, passive, or total force generated at a specific point in time.

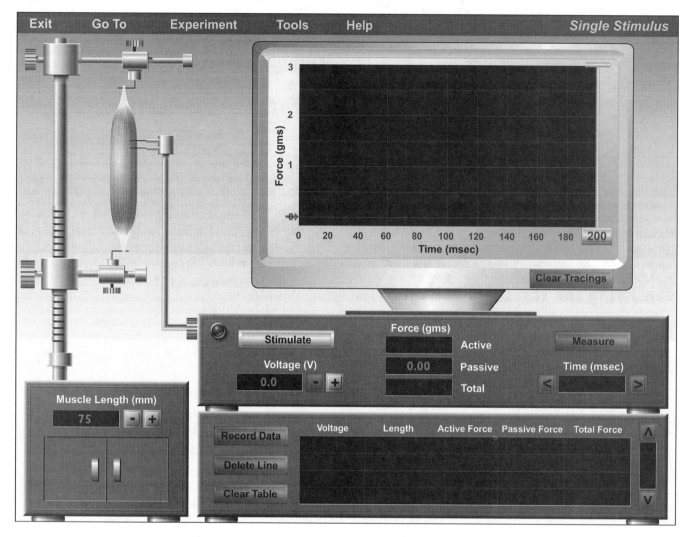

Figure 2.1 Opening screen of the Single Stimulus experiment.

Beneath the stimulator is the data collection box. Clicking on **Record Data** after an experimental run will allow you to record the data in this box. To delete a line of data, click on the data to highlight it and then click **Delete Line.** You may also delete the entire table by clicking **Clear Table.**

Activity 1:
Identifying the Latent Period

Recall that the **latent period** is the period of time that elapses between the generation of an action potential in a muscle cell and the start of muscle contraction.

1. Set the **Voltage** to 6.0 volts by clicking the (+) button on the stimulator until the voltage display reads 6.0.

2. Click **Stimulate** and observe the tracing that results. Notice that the trace starts at the left side of the screen and stays flat for a short period of time. Remember that the X axis displays elapsed time.

3. Click on the **Measure** button on the stimulator. Note that a thin, vertical yellow line appears at the far left side of the oscilloscope screen.

4. Click on the (>) button underneath **Time (msec).** You will see the vertical yellow line start to move across the screen. Watch what happens in the **Time (msec)** display as the line moves across the screen. Keep clicking the (>) button until the yellow line reaches the point in the tracing where the graph stops being a flat line and begins to rise (this is the point at which muscle contraction starts.) If the yellow line moves past the desired point, you can use the (<) button to move it backwards.

How long is the latent period? _____ msec

Note: If you wish to print your graph, click **Tools** on the menu bar and then click **Print Graph.**

5. Increase or decrease the stimulus voltage and repeat the experiment. (Remember that you can clear the tracings on the screen at any time by clicking **Clear Tracings.**) Record your data here:

Stimulus
voltage: _____ V

Latent
period: _____ msec

Stimulus
voltage: _____ V

Latent
period: _____ msec

Stimulus
voltage: _____ V

Latent
period: _____ msec

Does the latent period change with different stimulus

voltages? _____

After completing this experiment, click **Clear Tracings** to clear the oscilloscope screen of all tracings. ■

Activity 2:
Identifying the Threshold Voltage

By definition, the **threshold** is the minimal stimulus needed to cause a depolarization of the muscle plasma membrane (sarcolemma.) The threshold is the point at which sodium ions start to move into the cell (instead of out of the cell) to bring about the membrane depolarization.

1. Set the **Voltage** on the stimulator to 0.0 volts.

2. Click **Stimulate.** What do you see in the Active Force display?

_____

3. Click **Record Data.**

4. Increase the voltage to 0.1 volt, then click **Stimulate.** Observe the oscilloscope screen and the Active Force display (on the right side of the stimulator).

5. Click **Record Data.**

6. Repeat steps 4 and 5 until you see a number that is greater than 0.00 appear in the Active Force display.

7. Print out the graph(s) that you see on the oscilloscope screen by clicking **Tools** at the top of the screen and then selecting **Print Graph.**

What is the threshold voltage? _____ V

How does the graph generated at the threshold voltage differ from the graphs generated at voltages below the threshold?

_____

_____ ■

Activity 3:
Effect of Increases in Stimulus Intensity

In this activity we will examine how additional increases in stimulus intensity (such as additional increases in voltage) affect muscle response.

1. Set the voltage to 0.5 volts and click **Stimulate.** Then click **Record Data.**

2. Continue increasing the voltage by 0.5 volts and clicking **Stimulate** until you have reached 10.0 volts. Observe the **Active Force** display and click **Record Data** after each stimulation. Leave all of your tracings on the screen so that you can compare them to one another. If you wish, you may click **Tools** and then **Print Graph** to print your tracings.

3. Observe your tracings. How did the increases in voltage affect the peaks in the tracings?

_____

_____

How did the increases in voltage affect the amount of active force generated by the muscle?

_____

_____

What is the voltage beyond which there were no further increases in active force? Maximal voltage: _____ V

Why is there a maximal voltage? What has happened to the muscle at this voltage? Keep in mind that the muscle we are working with consists of many individual muscle fibers.

_____

_____

_____

An individual muscle fiber follows the *all-or-none* principle—it will either contract 100% or not at all. Does the muscle we are working with exhibit the *all-or-none* principle? Why or why not?

_____

_____

_____

_____

4. If you wish, you may view a summary of your data on a plotted data grid by clicking **Tools** and then **Plot Data.**

5. Click **Tools → Print Data** to print your data. ■

Multiple Stimulus

Click on **Experiment** at the top of the screen and then select **Multiple Stimulus.** You will see a slightly different screen appear (Figure 2.2.) The main change is that a **Multiple Stimulus** button has now been added to the electrical stimulator. This button allows you to start and stop the stimulator as you wish. When you click **Multiple Stimulus,** you'll notice that the button's label changes to **Stop Stimulus.** Clicking on **Stop Stimulus** turns off the stimulator.

Activity 4:

Treppe

Treppe is the progressive increase in force generated when a muscle is stimulated at a sufficiently high frequency. At such a frequency, muscle twitches follow one another closely, with each successive twitch peaking slightly higher than the one before. This step-like increase in force is why treppe is also known as the *staircase phenomenon.*

1. Set the voltage to the maximal voltage you established in Activity 3.

2. Click on the **200** button on the far right edge of the oscilloscope screen and slowly drag it as far to the left of the screen as it will go. This will allow you to see a longer span of time displayed on the screen.

3. Click **Single Stimulus** once. You will watch the trace rise and fall. As soon as it falls, click **Single Stimulus** again. Watch the trace rise and fall; when it falls, click **Single Stimulus** a third time.

What do you observe?

_____

_____

_____

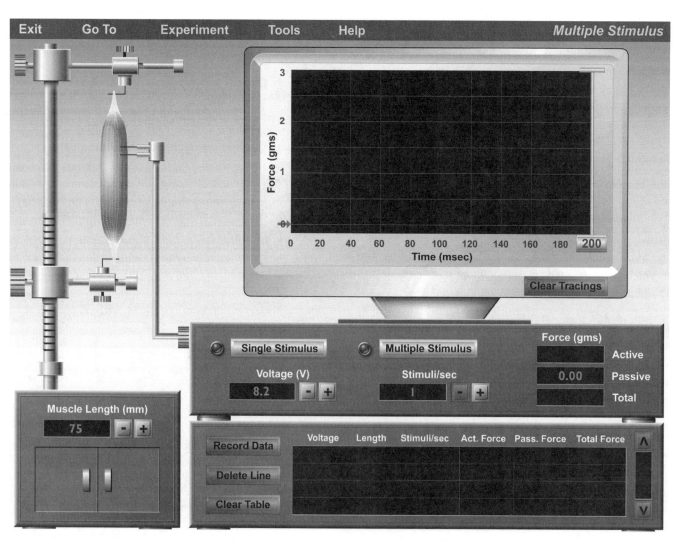

Figure 2.2 Opening screen of the Multiple Stimulus experiment.

4.   Click on the **200** button and drag it back to the far right edge of the oscilloscope screen.

5.   To print your graphs, click on **Tools** on the menu bar, then click **Print Graph.** ▪

Activity 5:
Summation

When a muscle is stimulated repeatedly, such that the stimuli arrive one after another within a short period of time, twitches can overlap with each other and result in a stronger muscle contraction than a stand-alone twitch. This phenomenon is known as **summation.** Summation occurs when muscle fibers that have already been stimulated once are stimulated again, before the fibers have relaxed.

1.   Set the **Voltage** to the maximal voltage you established in Activity 3.

2.   Click on the **Single Stimulus** button and observe the oscilloscope screen.

What is the active force of the contraction? _____ gms

3.   Click on the **Single Stimulus** button once. Watch the trace rise and begin to fall. Before the trace falls completely, click **Single Stimulus** again. (You may want to simply click **Single Stimulus** twice in quick succession in order to achieve this.)

What is the active force now? _____ gms.

4.   Click on **Single Stimulus** and allow the graph to rise and fall before clicking **Single Stimulus** again.

Was there any change in the force generated by the muscle?

_____

5.   Click on **Single Stimulus** and allow the graph to rise, but not fall, before clicking **Single Stimulus** again.

Was there any change in the force generated by the muscle?

_____

Why has the force changed?

_____

_____

_____

6.   Decrease the voltage on the stimulator and repeat this activity.

Do you see the same pattern of changes in the force

generated? _____

7. Next, stimulate the muscle as fast as you can (that is, click on **Single Stimulus** several times in quick succession).

Does the force generated change with each additional stimulus? If so, how?

_____

_____

_____ ▪

Activity 6:
Tetanus

In the previous activity we observed that if stimuli are applied to a muscle frequently in quick succession, the muscle generated more force with each successive stimulus. However, if stimuli continue to be applied frequently to a muscle over a prolonged period of time, the muscle force will eventually reach a plateau—a state known as **tetanus.** If stimuli are applied with even greater frequency, the twitches will begin to fuse so that the peaks and valleys of each twitch become indistinguishable from one another—this state is known as *complete (fused) tetanus.* The stimulus frequency at which no further increases in force are generated by the muscle is the *maximal tetanic tension.*

1.   Click **Clear Tracings** to erase any existing tracings from the oscilloscope screen.

2.   Underneath the **Multiple Stimulus** button, set the **Stimuli/sec** display, located beneath the **Multiple Stimulus** button, to 50 by clicking on the (**+**) button.

3.   Set the voltage to the maximal voltage you established in Activity 3.

4.   Click **Multiple Stimulus** and watch the trace as it moves across the screen. You will notice that the **Multiple Stimulus** button changes to a **Stop Stimulus** button as soon as it is clicked. After the trace has moved across the full screen and begins moving across the screen a second time, click the **Stop Stimulus** button.

What begins to happen at around 80 msec?

_____

What is this condition called?

_____

5.   Leave the trace on the screen. Increase the **Stimuli/sec** setting to 130 by clicking the (**+**) button. Then click **Multiple Stimulus** and observe the trace. After the trace has moved across the full screen and begins moving across the screen a second time, click **Stop Stimulus.**

How does the trace at 130 stimuli/sec compare with the trace at 50 stimuli/sec?

_____

_____

What is this condition called?

_____

_____

6. Click **Clear Tracings** to clear the oscilloscope screen.

7. Increase the **Stimuli/sec** setting to 145 by clicking the (+) button. Then click **Multiple Stimulus** and observe the trace. Click **Stop Stimulus** after the trace has swept one full screen. Then click **Record Data.**

8. Repeat step 7, increasing the **Stimuli/sec** setting by 1 each time until you reach 150 stimuli/sec. (That is, set the **Stimuli/sec** setting to 146, then 147, 148, etc.) Be sure to click **Record Data** after each run.

9. Examine your data. At what stimulus frequency is there no further increase in force?

_____

_____

What is this stimulus frequency called?

_____

_____

10. For another view of your data, click **Tools** and then click **Plot Data.**

11. Click **Clear Tracings** to clear the oscilloscope screen. If you wish to print your data, click **Tools** and then **Print Data.** ■

Activity 7:
Fatigue

Fatigue is a decline in a muscle's ability to maintain a constant force of contraction after prolonged, repetitive stimulation. The causes of fatigue are still being investigated, though in the case of high-intensity exercise, lactic acid buildup in muscles is thought to be a factor. In low-intensity exercise, fatigue may be due to a depletion of energy reserves.

1. Design an experiment that demonstrates fatigue on the oscilloscope screen. Hint: Set **Stimuli/sec** to above 100. In fatigue, what happens to force production over time?

_____

_____

2. Print out your results by clicking on **Tools**, then **Print Graph.**

3. Click **Tools** → **Print Data** to print your recorded data. ■

Isometric and Isotonic Contractions

Muscle contractions can be either **isometric** or **isotonic.** When a muscle attempts to move a load that is greater than the force generated by the muscle, the muscle contracts _isometrically_. In this type of contraction, the muscle stays at a fixed length (isometric means "same length"). An example of isometric muscle contraction is when you stand in a doorway and push on the doorframe. The load that you are attempting to move (the doorframe) is greater than the force generated by your muscle, and so your muscle does not shorten.

When a muscle attempts to move a load that is equal to or less than the force generated by the muscle, the muscle contracts _isotonically_. In this type of contraction, the muscle shortens during a period of time in which the force generated by the muscle remains constant (isotonic means "same tension"). An example of isotonic contraction is when you lift a book from a tabletop. The load that you are lifting (the book) is equal to or less than the force generated by your muscle. Your muscle shortens when it contracts, allowing you to lift the book.

We will first examine isometric contraction. Click on **Experiment** at the top of the screen and then select **Isometric Contraction.** You will see the screen shown in Figure 2.3. Notice that there are now two oscilloscope screens. The screen on the left is basically identical to the ones you worked with in the previous activities. The screen on the right is new. The Y axis is still "Force," but the X axis is now muscle length. Let's conduct a practice run to familiarize ourselves with the equipment.

1. Leave the **Voltage** set at 8.2 and click **Stimulate.** You will see indicators of three forces in the right-hand screen. You will see the Passive Force indicator near the bottom of the screen (in green), and the Active Force (in purple) and Total Force (in yellow) indicators together higher up on the screen. Note that the Active Force indicator is seen inside of the Total Force indicator.

2. Click the (+) and (−) buttons beneath **Muscle Length** on the left side of the screen and notice how the muscle may be stretched or shortened.

3. You are now ready to begin the experiment. Click the **Clear Tracings** and **Clear Plot** buttons under each oscilloscope screen. ■

Activity 8:
Isometric Contractions

1. Leave the **Voltage** set at 8.2.

2. On the lower left side of the screen, click the (−) button underneath **Muscle Length** and reduce the length to 50 mm.

3. Click **Stimulate** and observe the results on both oscilloscope screens.

4. Click **Record Data** at the bottom of the screen.

5. Repeat steps 1–5, increasing the muscle length by 10 mm each time (i.e., 60 mm, then 70 mm, then 80 mm, etc.) until you reach 100 mm. Remember to click **Record Data** after each run.

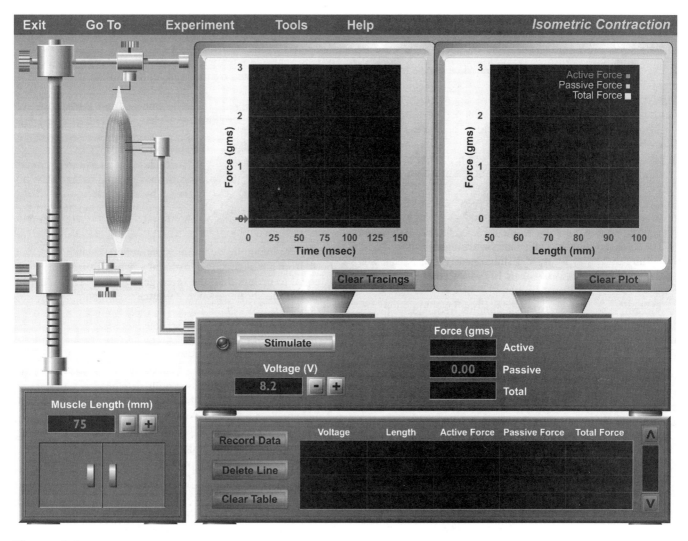

Figure 2.3 Opening screen of the Isometric Contraction experiment.

6.   Click **Tools** at the top of the screen, then click **Plot Data.** You will see a screen pop up depicting a plotted graph. Be sure that *Length* is depicted on the X axis and *Active Force* is depicted on the Y axis. You may wish to click **Print Plot** at the top left corner of the window to print the graph.

Looking at this graph, what muscle lengths generated the most active force? (provide a range)

_____-to-_____ mm

At what muscle length does passive force begin to play less of a role in the total force generated by the muscle?

_____ mm

7.   Move the blue square bar for the Y axis to *Passive Force.* You may wish to click **Print Plot** at the top left corner of the window to print the graph.

Looking at this graph, at what muscle length does Passive

Force begin to pay a role in the total force generated by the muscle?

_____ mm

8.   Move the blue square bar for the Y axis to *Total Force.* You may wish to click on **Print Plot** at the top left corner of the window to print the graph.

9.   Click **Tools → Print Data** to print your data.

The graph shows a dip at muscle length = 90 mm. Why is this?

_____

_____

_____

What is the key variable in an isometric contraction?

_____

Activity 9:
Isotonic Contractions

Recall that isotonic contraction occurs when a muscle generates a force equal to or greater than the load it is opposing. In this type of contraction, there is a latent period, followed by a rise in generation of force, followed by a period of time during which the force produced by the muscle remains constant (recall that isotonic means "same tension.") During this plateau period, the muscle shortens and is able to move the load. The muscle is not able to shorten prior to the plateau because enough force has not yet been generated to move the load. When the force generated becomes equal to the load, the muscle shortens. The force generated will be constant so long as the load is moving. Eventually, the muscle will relax and the load will begin to fall. An isotonic twitch is not an all-or-nothing event. If the load is increased, the muscle must generate more force to move it. The latent period will also lengthen, as it will take more time for the necessary force to be generated by the muscle. The speed of the contraction depends on the load the muscle is opposing. Maximal speed is attained with minimal load. Conversely, the heavier the load, the slower the muscle contraction.

Click on **Experiment** at the top of the screen and then select **Isotonic Contraction.** The screen that appears (Figure 2.4) is similar to the *Single Stimulus* screen you worked with in Activities 1–3. Note that fields for Muscle Length and Velocity have been added to the display below the oscilloscope screen, and that the muscle on the left side of the screen is now dangling freely at its lower end. The weight cabinet below the muscle is open; inside are four weights, each of which may be attached to the muscle. Above the weight cabinet is a moveable platform, which you may move by clicking the (+) or (−) buttons under **Platform Height.**

In this experiment, you will be attaching weights to the end of the muscle in order to observe isotonic contraction.

1. The **Voltage** should already be set at 8.2, and the **Platform Height** at 75 mm. If not, adjust the settings accordingly.

2. Click on the **0.5g** weight in the weight cabinet and attach it to the dangling end of the muscle. The weight will pull down on the muscle and come to rest on the platform.

3. Click **Stimulate** and observe the trace. Note the rise in force, followed by a short plateau, followed by a relaxation phase. Note that the Active Force display is the same as the weight that was attached: 0.5 grams.

How much time does it take for the muscle to generate

0.5 grams of force? _____ msec

4. Click **Stimulate** again, watching the muscle and the screen at the same time as best you can. Then click **Record Data.**

At what point in the trace does the muscle shorten?

_____

You can observe from the trace that the muscle is rising in force before it reaches the plateau phase. Why doesn't the muscle shorten prior to the plateau phase?

_____

_____

5. Remove the .5g weight and attach the **1.0g** weight. Leave your previous trace on the screen.

6. Click **Stimulate** and then **Record Data.**

Did it take any longer for the muscle to reach the force it needed to move the weight?

_____

How does this trace differ from the trace you generated with the .5g weight attached?

_____

7. Leaving the two previous traces on the screen, repeat the experiment with the two remaining weights. Click **Record Data** after each result. If you wish to print your graphs, click **Tools** and then select **Print Graph.**

8. When you have finished recording data for all four weights, click **Tools** from the top of the screen and click **Plot Data.**

9. Move the blue square bar for the Y axis to *Velocity* and the blue square bar for the X axis to *Weight.*

Examine the plot data and your numerical data. At what weight was the velocity of contraction the fastest?

_____ gms

What happened when you attached the 2.0g weight to the muscle and stimulated the muscle? How did this trace differ from the other traces? What kind of contraction did you observe?

_____

_____

_____

10. Close the Plot Data window by clicking on the "X" at the top right corner of the window. If you still have a weight attached to the muscle, remove it. Also click **Clear Tracings** to clear the oscilloscope screen.

11. Place the **0.5g** weight on the muscle and adjust the height of the platform to 100 mm. Notice that the platform no longer supports the weight.

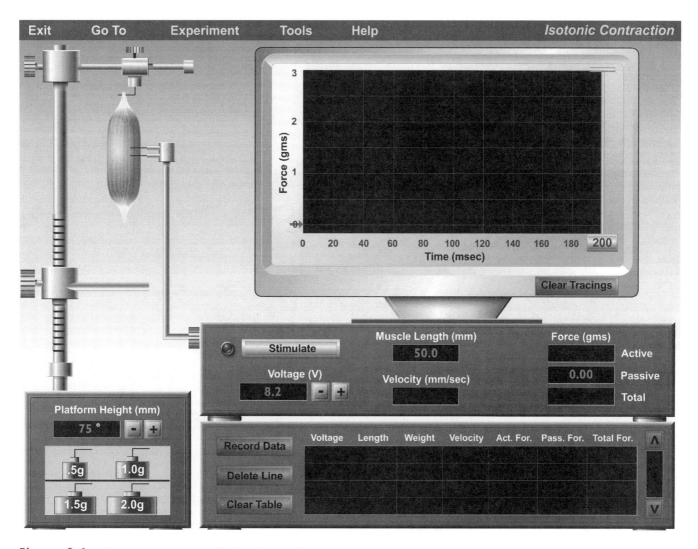

Figure 2.4 Opening screen of the Isotonic Contraction experiment.

12. Click **Stimulate** and observe the muscle contraction tracing. What kind of a trace did you get?

_____

What was the force of the contraction? _____ gms

13. Click **Record Data.** Then repeat steps 12–13 for each of the remaining weights (remember to click **Record Data** after obtaining the results for each weight). Click **Tools** and then **Print Graph** if you wish to print your tracings.

Describe your four tracings and explain what has happened in each of them.

_____

_____

_____

_____

_____

_____

_____

14. Click **Clear Tracings.**

15. Place the **1.5g** weight on the muscle.

16. Adjust the platform height to 90 mm.

17. Click **Stimulate** and then **Record Data.**

18. Repeat steps 16–18 except lower the platform height by 10 mm each time until you reach 60 mm. (i.e., set the platform height at 80 mm, then 70 mm, then 60 mm)

19. Click **Tools,** then **Plot Data.**

20. Within the Plot Data window, move the blue square bar for the X axis to *Length* and the blue square bar for the Y axis to *Velocity.*

What muscle length(s) generated the fastest contraction velocity?

_____

22. Close the Plot Data window by clicking the "X" in the upper right corner of the window.

23. Click **Tools** → **Print Data** to print your data.

Histology Review Supplement

Turn to p. 136 for a review of skeletal muscle tissue.

Neurophysiology of Nerve Impulses

Objectives

1. To define the following: irritability, conductivity, resting membrane potential, polarized, sodium-potassium pump, threshold stimulus, depolarization, action potential, repolarization, hyperpolarization, absolute refractory period, relative refractory period, nerve impulse, synaptic cleft, compound action potential, conduction velocity
2. To list at least four different stimuli capable of generating an action potential
3. To list at least two agents capable of inhibiting an action potential
4. To describe the relationship between nerve size and conduction velocity
5. To describe the relationship between nerve myelination and conduction velocity

The nervous system is responsible for most of the functions that characterize higher organisms, such as muscular movement, awareness, thought, learning, and memory. **Neurons** are the functional cellular units of the nervous system. They are "excitable" cells that communicate by transmitting electrical impulses ("excitable" means that they are capable of producing large, rapid electrical signals called **action potentials**). Neurons are specialized for receiving, integrating, and transmitting information to other neurons and/or effector cells. A typical neuron consists of a *cell body,* containing its nucleus and organelles; *dendrite(s),* responsible for carrying nerve impulses toward the cell body; and an *axon,* responsible for carrying nerve impulses away from the cell. Junctions between cells are called *synapses,* where one cell (the *presynaptic* cell) releases a chemical messenger called a *neurotransmitter* that communicates with the dendrite or cell body of a *postsynaptic* cell. While synaptic transmission is usually thought of as being excitatory (inciting an action potential in the postsynaptic cell), some are *inhibitory.* This is accomplished by causing the postsynaptic cell to become *hyperpolarized,* or having a resting membrane potential that is more negative than the normal resting membrane potential.

Neurons have two major physiologic properties: **irritability,** the ability to respond to stimuli and convert them into nerve impulses, and **conductivity,** the ability to transmit an impulse (in this case, to take the neural impulse and pass it along the cell membrane). In the resting neuron (that is, a neuron that is neither receiving nor transmitting any signals), the exterior of the cell membrane is positively charged and the interior of the neuron is negatively charged. This difference in electrical charge across the plasma membrane is referred to as the **resting membrane potential,** and the membrane is said to be **polarized.** The **sodium-potassium pump** in the membrane maintains the difference in electrical charge established by diffusion of ions. This active transport mechanism moves three sodium ions out of the cell while moving in two potassium ions. Therefore, the major cation in the extracellular fluid outside of the cell is sodium, while the major cation inside of the cell is potassium. The inner surface of the cell membrane is more negative than the outer surface, mainly due to **intracellular proteins,** which, at body pH, tend to be negatively charged.

The resting membrane potential can be measured with a voltmeter by putting a recording electrode just inside the cell membrane and by placing a reference, or ground, electrode outside of the membrane. In the giant squid axon (where most early neural research was conducted), and in the frog axon that will be used in this exercise, the resting membrane potential is −70 mv. (In humans, the resting membrane potential typically measures between −40 mv to −90 mv).

When a neuron is activated by a stimulus of sufficient intensity, known as a **threshold stimulus,** the membrane at its *trigger zone,* typically the axon hillock, briefly becomes more permeable to sodium ions (sodium gates in the cell membrane open). Sodium ions rush into the cell, increasing the number of positive ions inside of the cell and changing the membrane polarity. The interior surface of the membrane becomes less negative and the exterior surface becomes less positive, a phenomenon called **depolarization.** When depolarization reaches a certain threshold, an action potential is initiated and the polarity of the membrane reverses.

When the membrane depolarizes, the resting membrane potential of −70 mv becomes less negative. When the membrane potential reaches 0 mv, indicating that there is no charge difference across the membrane, the sodium ion channels close and potassium ion channels open. By the time the sodium ions channels finally close, the membrane potential has reached +35 mv. The opening of the potassium ion channels allows potassium ions to flow out of the cell down their electrochemical gradient—remember that like ions are repelled from each other. The flow of potassium ions out of the cell causes the membrane potential to move in a negative direction. This is referred to as **repolarization.** This repolarization occurs within a millisecond of the initial sodium influx and reestablishes the resting membrane potential. Actually, by the time the potassium gates close, the cell membrane has undergone a hyperpolarization, slipping to perhaps −75 mv. With the gates closed, the resting membrane potential is quickly returned to the normal resting membrane potential.

When the sodium gates are open, the membrane is totally insensitive to additional stimuli, regardless of their force. The cell is in what is called the **absolute refractory period.** During repolarization, the membrane may be stimulated by a very strong stimulus. This period is called the **relative refractory period.**

The action potential, once started, is a self-propagating phenomenon, spreading rapidly along the neuron membrane. The action potential follows the *all-or-none* law, in which the neuron membrane either depolarizes 100% or not at all. In neurons, the action potential is also called a **nerve impulse.** When it reaches the **axon terminal,** it triggers the release of neurotransmitters into a gap, known as the **synaptic cleft.** Depending on the situation, the neurotransmitter will either excite or inhibit the postsynaptic neuron.

In order to study nerve physiology, we will use a frog nerve and several electronic instruments. The first instrument we will use is an **electronic stimulator.** Nerves can be stimulated by chemicals, touch, or electric shock. The electronic stimulator administers an electric shock that is pure DC, and it allows the duration, frequency, and voltage of the shock to be precisely controlled. The stimulator has two output terminals; the positive terminal is red and the negative terminal is black. Voltage leaves the stimulator via the red terminal, passes through the item to be stimulated (in this case, the nerve), and returns to the stimulator at the black terminal to complete the circuit.

The second instrument is an **oscilloscope,** an instrument that measures voltage changes over a period of time. The face of the oscilloscope is similar to a black-and-white TV screen. The screen of the oscilloscope is the front of a tube with a filament at the back. The filament is heated and gives off a beam of electrons that passes to the front of the tube. Electronic circuitry allows for the electron beam to be brought across the screen at preset time intervals. When the electrons hit the phosphorescent material on the inside of the screen, a spot on the screen will glow. When we apply a stimulus to a nerve, the oscilloscope screen will display one of the following three results: no response, a flat line, or a graph with a peak. A graph with a peak indicates that an action potential has been generated.

While performing the following experiments, keep in mind that you are working with a **nerve,** which consists of many neurons—you are not just working with a single neuron. The action potential you will see on the oscilloscope screen reflects the cumulative action potentials of all the neurons in the nerve, called a **compound nerve action potential.** Although an action potential follows the *all-or-none* law within a single neuron, it does not necessarily follow this law within an entire nerve. When you electrically stimulate a nerve at a given voltage, the stimulus may result in the depolarization of most of the neurons, but not necessarily all of them. To achieve depolarization of *all* of the neurons, a higher stimulus voltage may be needed.

Eliciting a Nerve Impulse

The **excitability** of a neuron—its ability to generate action potentials—is what allows neurons to perform their functions. In the following experiments you will be investigating

what kinds of stimuli trigger an action potential. To begin, follow the instructions for starting PhysioEx in the "Getting Started" section at the front of this manual. From the main menu, select **Neurophysiology of Nerve Impulses.** The opening screen will appear in a few seconds (see Figure 3.1). Note that a sciatic nerve from a frog has been placed into the nerve chamber. Leads go from the stimulator output to the nerve chamber, and also go from the nerve chamber to the oscilloscope. Notice that these leads are red and black. The stimulus travels along the red lead to the nerve. When the nerve depolarizes, it will generate an electrical impulse that will travel along the red wire to the oscilloscope, and back to the nerve along the black wire.

A c t i v i t y   1 :
Electrical Stimulation

1.   Set the voltage at 1.0 V by clicking the (**+**) button next to the **Voltage** display.

2.   Click **Single Stimulus.**

Do you see any kind of response on the oscilloscope screen?

_____

If you saw no response, or a flat line indicating no action potential, click the **Clear** button on the oscilloscope, increase the voltage, and click **Single Stimulus** again until you see a trace (deflection of the line) that indicates an action potential.

What was the *threshold voltage,* or the voltage at which you

first saw an action potential? _____ V

Click **Record Data** on the data collection box to record your results.

3.   If you wish to print your graph, click **Tools** and then **Print Graph.** You may do this each time after you have generated a graph on the oscilloscope screen.

4.   Increase the voltage by 0.5 V and click **Single Stimulus.**

How does this tracing compare to the one that was generated at the threshold voltage? (Hint: Look very carefully at the tracings.)

_____

What reason can you give for your answer?

_____

_____

_____

Click **Record Data** on the data collection box to record your results.

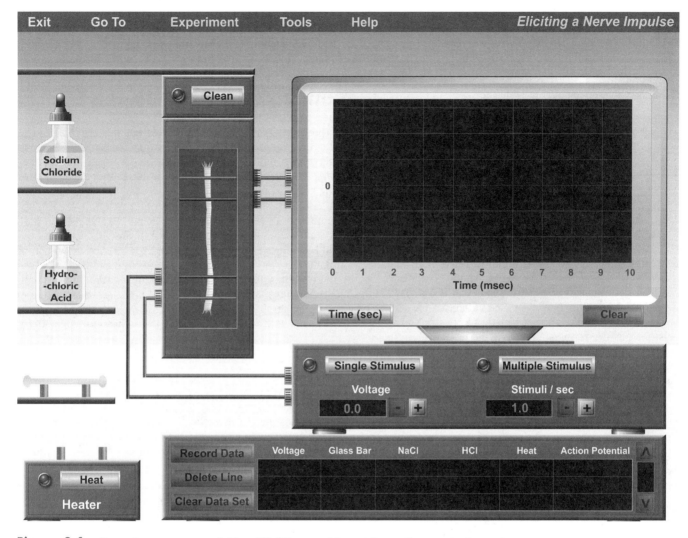

Figure 3.1 Opening screen of the Eliciting a Nerve Impulse experiment.

5.   Continue increasing the voltage by 0.5 V and clicking **Single Stimulus** until you find the point beyond which no further increase occurs in the peak of the action potential trace.

Record this maximal voltage here: _____ V

Click **Record Data** to record your results.

Now that you have seen that an electrical impulse can cause an action potential, let's try some other methods of stimulating a nerve. ■

Activity 2:
Mechanical Stimulation

1.   Click the **Clear** button on the oscilloscope.

2.   Using the mouse, click and drag the glass bar to the nerve and place it over the nerve. When the glass rod is touching the nerve, release the mouse button. What do you see on the oscilloscope screen?

How does this tracing compare with the other tracings you have generated?

_____

Click **Record Data** to record your results. Leave the graph on the screen so that you can compare it to the graph you will generate in the next activity. ■

Activity 3:
Thermal Stimulation

1.   Click on the glass rod and drag it to the heater, and then release the mouse button. Click the **Heat** button. When the rod turns red, indicating that it has been heated, click and drag the rod over the nerve and release the mouse button. What happens?

_____

How does this trace compare to the trace that was generated with the unheated glass bar?

_____

_____

What explanation can you provide for this?

_____

_____

Click **Record Data** to record your results. Then click **Clear** to clear the oscilloscope screen for the next activity. ∎

Activity 4:
Chemical Stimulation

1. Click and drag the dropper from the bottle of sodium chloride (salt solution) over to the nerve in the chamber and then release the mouse button to dispense drops. Does this generate an action potential?

_____

_____

2. Using your threshold voltage setting, stimulate the nerve. Does this tracing differ from the original threshold stimulus tracing? If so, how?

_____

_____

Click **Record Data** to record your results.

3. Click the **Clean** button on top of the nerve chamber. This will return the nerve to its original (non-salted) state. Click **Clear** to clear the oscilloscope screen.

4. Click and drag the dropper from the bottle of hydrochloric acid over to the nerve, and release the mouse button to dispense drops. Does this generate an action potential?

_____

_____

5. Does this tracing differ from the one generated by the original threshold stimulus?

_____

_____

Click **Record Data** to record your results.

6. Click the **Clean** button on the nerve chamber to clean the chamber and return the nerve to its untouched state.

7. Click **Tools → Print Data** to print your data.

To summarize your experimental results, what kinds of stimuli can elicit an action potential?

_____

_____

_____

You have reached the end of this activity. To continue on to the next activity, click the **Experiment** pull-down menu and select **Inhibiting a Nerve Impulse.** ∎

Inhibiting a Nerve Impulse

The local environment of most neurons is controlled so that the neurons are protected from changes in the composition of the interstitial fluid. However, numerous physical factors and chemical agents can impair the ability of nerve fibers to function. For example, deep pressure and cold temperature both block nerve impulse transmission by preventing local blood supply from reaching the nerve fibers. Local anesthetics, alcohol, and numerous other chemicals are also very effective in blocking nerve transmission. In this activity we will study the effects of various agents on nerve transmission.

The display screen for this activity is very similar to the screen in the first activity (Figure 3.2). To the left are bottles of three agents that we will test on the nerve. Keep the tracings you printed out from the first activity close at hand for comparison.

Activity 5:
Testing the Effects of Ether

1. Using the mouse, click and drag the dropper from the bottle marked Ether over to the nerve, in between the stimulating electrodes and recording electrodes. Release the mouse button to dispense drops.

2. Click **Stimulate,** using the voltage setting from the threshold stimulus you used in the earlier activities. What sort of trace do you see?

_____

_____

What has happened to the nerve?

_____

_____

Click **Record Data** to record your results.

3. Click the **Time (min.)** button on the oscilloscope. The screen will now display activity over the course of 10 minutes (the space between each vertical line represents 1 minute each). Because of the change in time scale, an action potential will look like a sharp vertical spike on the screen.

4. Click the (+) button under **Interval between Stimuli** on the stimulator to set the interval to 2.0 minutes. This will set the stimulus to stimulate the nerve every two minutes. Click on **Stimulate** to start the stimulations. Watch the **Elapsed Time** display.

How long does it take for the nerve to return to normal?

_____

_____

5. Click the **Stop** button to stop this action and to return the elapsed time to 0.00.

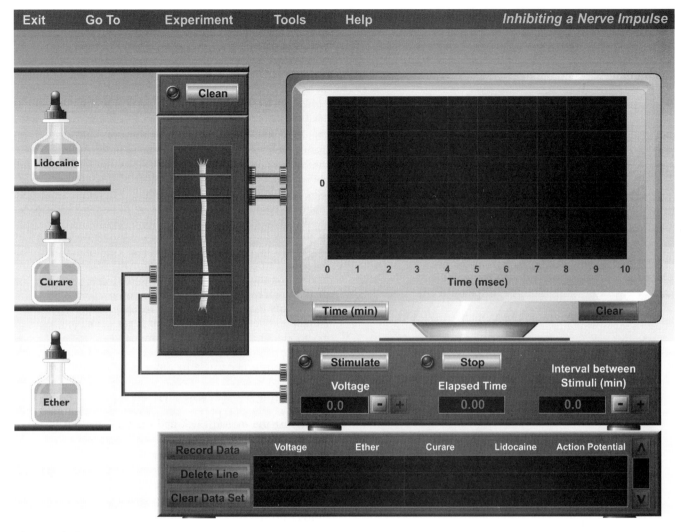

Figure 3.2 Opening screen of the Inhibiting a Nerve Impulse experiment.

6. Click the **Time (msec)** button on the oscilloscope to re-turn it to its normal millisecond display.

7. Click **Clear** to clear the oscilloscope for the next activity.

8. Click the (−) button under **Interval between Stimuli** until it is reset to 0.0.

9. Click the **Clean** button on the nerve chamber to clean the chamber and return the nerve to its untouched state.

Activity 6:
Testing the Effects of Curare

Curare is a well-known plant extract that South American Indians used to paralyze their prey. It is an alpha-toxin that binds to acetylcholine binding sites on the postsynaptic cell membrane, which prevents the acetylcholine from acting. Curare blocks synaptic transmission by preventing neural impulses to flow from neuron to neuron.

1. Click and drag the dropper from the bottle marked Curare, and position the dropper on the nerve, between the stimulating and recording electrodes. Release the mouse button to dispense drops.

2. Set the stimulator at the threshold voltage and stimulate the nerve. What effect on the action potential is noted?

_____

What explains this effect?

_____

What do you think would be the overall effect of Curare on the organism?

_____

_____

Click **Record Data** to record your results.

3. Click the **Clean** button on the nerve chamber to remove the curare and return the nerve to its original untouched state.

4. Click **Clear** to clear the oscilloscope screen for the next activity. ■

Activity 7:
Testing the Effects of Lidocaine

Note: Lidocaine is a sodium-channel antagonist.

1. Click and drag the dropper from the bottle marked Lidocaine, and position it over the nerve, between the stimulating and recording electrodes. Release the mouse button to dispense drops. Does this generate a trace?

_____

2. Stimulate the nerve at the threshold voltage. What sort of tracing is seen?

_____

Why does lidocaine have this effect on nerve fiber transmission?

_____

_____

Click **Record Data** to record your results.

3. Click the **Clean** button on the nerve chamber to remove the lidocaine and return the nerve to its original untouched state.

4. Click **Tools → Print Data** to print your data.

You have reached the end of this activity. To continue on to the next activity, click the **Experiment** pull-down menu and select **Nerve Conduction Velocity**. ■

Nerve Conduction Velocity

The speed of transmission of information in the nervous system depends in part on the conduction velocity of the axon, carrying information from the cell body to the next cell. Conduction velocity depends on the size, or diameter, of the axon, and whether or not it is myelinated. As has been pointed out, one of the major physiological properties of neurons is **conductivity:** the ability to transmit the nerve impulse to other neurons, muscles, or glands. The nerve impulse, or propagated action potential, occurs when sodium ions flood into the neuron, causing the membrane to depolarize. Although this event is spoken of in electrical terms and is measured using instruments that measure electrical events, the velocity of the action potential along a neural membrane does not occur at the speed of light. Rather, this event is much slower. In certain nerves in the human, the velocity of an action potential may be as fast as 120 meters per second. In other nerves, conduction speed is much slower, occurring at a speed of less than 3 meters per second.

In this exercise, the oscilloscope and stimulator will be used along with a third instrument, the **bio-amplifier.** The bio-amplifier is used to amplify any membrane depolarization so that the oscilloscope can easily record the event. Normally, when a membrane depolarization sufficient to initiate an action potential occurs, the interior of the cell membrane goes from -70 millivolts to about $+40$ millivolts. This is easily registered and viewable on an oscilloscope, without the aid of an amplifier. However, in this experiment, it is the change in the membrane potential on the *outside* of the nerve that is being observed. The change that occurs here during depolarization is so miniscule that it must be amplified in order to be visible on the oscilloscope.

A nerve chamber (similar to the one used in the previous two experiments) will be used. Basically, this is a plastic box with platinum electrodes running across it. The nerve will be laid on these electrodes. Two electrodes will be used to bring the impulse from the stimulator to the nerve, and three electrodes will be used for recording the membrane depolarization.

In this experiment we will measure and compare the conduction velocities of different types of nerves. We will examine four nerves: an earthworm nerve, a frog nerve, and two rat nerves. The earthworm nerve is the smallest of the four. The frog nerve is a medium-sized **myelinated** nerve (consult your text's discussion of myelination). Rat nerve #1 is a medium-sized **unmyelinated** nerve. Rat nerve #2 is a large, myelinated nerve—the largest nerve in this group. We will observe the effects of size and myelination on nerve conductivity.

The basic layout of the materials is shown in Figure 3.3. The two wires (red and black) from the stimulator connect with the top right side of the nerve chamber. Three wires (red, black, and a bare wire cable) are attached to connectors on the bottom left side of the nerve chamber and go to the bio-amplifier. The bare cable serves as a "ground reference" for the electrical circuit and provides the reference for comparison of any change in membrane potential. The bio-amplifier is connected to the oscilloscope so that any amplified membrane changes can be observed. The stimulator output, called the "pulse," has been connected to the oscilloscope so that when the nerve is stimulated, the tracing will start across the oscilloscope screen. Thus, the time from the start of the trace on the left-hand side of the screen (when the nerve was stimulated) to the actual nerve deflection (from the recording electrodes) can be accurately measured. This amount of time, usually measured in milliseconds, is critical for determining conduction velocity.

Look closely at the screen. The wiring of the circuit may seem complicated, but it really is not. First, look at the stimulator, located on top of the oscilloscope. On the left side, red and black wires leave the stimulator to go to the nerve chamber. Remember, the red wire is the "hot" wire that carries the impulse from the stimulator, and the black wire is the return to the stimulator that completes the circuit. When the nerve is stimulated, the red recording wire (leaving the left side of the nerve chamber) will pick up the membrane impulse and bring it to the bio-amplifier. The black wire, as before, completes the circuit, and the bare cable wire simply acts as a reference electrode. The membrane potential, picked up by the red wire, is then amplified by the bio-amplifier, and the output is carried to the oscilloscope. The oscilloscope then shows the trace of the nerve action potential.

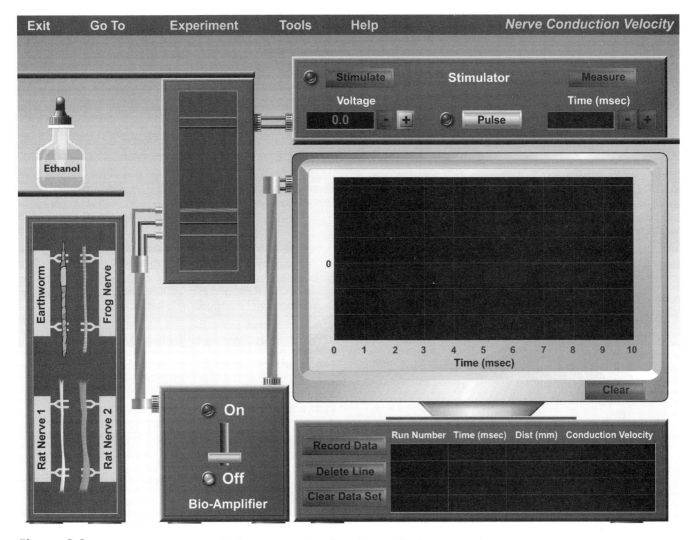

Figure 3.3   Opening screen of the Nerve Conduction Velocity experiment.

A c t i v i t y   8 :
Measuring Nerve Conduction Velocity

1.   On the stimulator, click on the **Pulse** button.

2.   Turn the bio-amplifier on by clicking the horizontal bar on the bio-amplifier and dragging it to the **On** setting.

On the left side of the screen are the four nerves that will be studied. The entire earthworm is used because it has a nerve running down its ventral surface. A frog nerve is used because the frog has long been the animal of choice in many physiology laboratories. The rat nerves are used so that you may compare a) the conduction velocity of different sized nerves, and b) the conduction velocity of a myelinated versus an unmyelinated nerve. Remember that the frog nerve is myelinated, and that rat nerve #1 is the same size as the frog nerve but unmyelinated. Rat nerve #2, the largest nerve of the bunch, is myelinated.

3.   Using the mouse, click and drag the dropper from the bottle labeled Ethanol over to the earthworm and release the mouse button to dispense drops of ethanol. This will narcotize the worm so it does not move around during the experiment, but will it not affect nerve conduction velocity. The alcohol is

at a low enough concentration that the worm will be fine and back to normal within 15 minutes.

4.   Click and drag the earthworm into the nerve chamber. Be sure the worm is over both of the stimulating electrodes and all three of the recording electrodes.

5.   Using the (+) button next to the **Voltage** display, set the voltage to 1.0 V. Then click **Stimulate** to stimulate the nerve.

Do you see an action potential? If not, increase the voltage by increments of 1.0 V until a trace is obtained. At what threshold voltage do you first see an action potential generated?

_____ V

6.   Next, click the **Measure** button located on the stimulator. You will see a vertical yellow line appear on the far left edge of the oscilloscope screen. Now click the (+) button under the Measure button. This will move the yellow line to the right. This line lets you measure how much time has elapsed on the graph at the point that the line is crossing the graph. You will see the elapsed time appear on the **Time (msec)**

display on the stimulator. Keep clicking (+) until the yellow line is right at the point in the graph where the graph ceases being a flat line and first starts to rise.

7. Once you have the yellow line positioned at the start of the graph's ascent, note the time elapsed at this point. Click **Record Data** to record this time on the data collection graph. PhysioEx will automatically compute the conduction velocity for you, based on this data. Note that the data collection box includes a **Distance (mm)** column, and that this distance is always 43 mm. This is the distance between the red stimulating wire and the red recording wire. In a wet lab, you would have to measure this distance yourself before you could proceed with calculating the conduction velocity.

It is very important that you have the yellow vertical measuring line positioned at the start of the graph's rise before you click **Record Data**—otherwise the conduction velocity calculated for the nerve will be inaccurate.

8. Fill in the data under the Earthworm column in the chart below:

9. Click and drag the earthworm to its original place. Click **Clear** to clear the oscilloscope screen.

10. Repeat steps 4–9 for the remaining nerves. Remember to click **Record Data** after each experimental run and to fill in the chart below.

11. Click **Tools → Print Data** to print your data.

Which nerve in the group has the slowest conduction velocity?

_____

What was the speed of the nerve? _____

Which nerve of the four has the fastest conduction velocity?

_____

What was the speed of the nerve? _____

What is the relationship between nerve size and conduction velocity? What are the physiological reasons for this relationship?

_____

_____

_____

_____

_____

Based on the results, what is your conclusion regarding the effects of myelination on conduction velocity? What are the physiological reasons for your conclusion?

_____

_____

_____

_____

What are the evolutionary advantages achieved by the myelination of neurons?

_____

_____

Histology Review Supplement

Turn to p. 137 for a review of nervous tissue.

Nerve	Earthworm (small nerve)	Frog (medium nerve, myelinated)	Rat Nerve #1 (medium nerve, unmyelinated)	Rat Nerve #2 (large nerve, myelinated)
Threshold voltage				
Elapsed time from stimulation to action potential				
Conduction velocity				

Endocrine System Physiology

Objectives

1. To define the following: hormones, target cell, negative feedback, metabolism, thyroxine, thyroid stimulating hormone (TSH), thyrotropin releasing hormone (TRH), hypothalamus, hypothalamic pituitary portal system, portal vein, hormone replacement therapy, diabetes type I, diabetes type II, glucose standard curve

2. To give examples of how negative feedback loops regulate hormone release

3. To explain the role of thyroxine in maintaining an animal's metabolic rate

4. To explain the effects of thyroid stimulating hormone (TSH) on an animal's metabolic rate

5. To understand the role of the hypothalamus in the regulation of thyroxine and TSH production

6. To understand how hypothalamic hormones reach the pituitary gland

7. To understand how estrogen affects uterine tissue growth

8. To explain how hormone replacement therapy works

9. To explain why insulin is important and how it can be used to treat diabetes

The endocrine system regulates the functioning of every cell, tissue, and organ in the body. It acts to maintain a stable internal body environment, regardless of changes occurring within or outside of the body. Endocrine cells have the ability to sense and respond to changes via the excretion of specific chemicals known as **hormones.** Hormones are carried in the blood, usually attached to specific plasma proteins, and circulate around the body. When the hormone-protein complex reaches a **target cell** (the cell at which a chemical message is aimed), the hormone detaches from the protein and enters the cell to induce a specific reaction.

Hormones work in different ways, depending upon their chemical structures. For example, *polypeptide hormones,* composed of chains of amino acids, work by first attaching to a protein receptor in the cell membrane, initiating a series of reactions in the membrane, resulting in *cyclic adenosine monophosphate* (cAMP) entering the cell. The entrance of this chemical into the cell induces the cell to work harder and faster. *Steroid hormones* and *thyroxine* (a hormone secreted by the thyroid, which we will be examining in detail shortly) enter the cell to attach to a cytoplasmic receptor. The hormone-receptor complex then enters the nucleus of the cell to attach to specific points on the DNA. Each attachment causes the production of a specific mRNA, which then moves to the cytoplasm to be translated into a specific protein.

Most regulation of hormone levels in the body is conducted by **negative feedback:** if a particular hormone is needed, production of that hormone will be stimulated; if there is enough of a particular hormone present, production of that hormone will be inhibited. In a few very specific instances, hormonal output is controlled by *positive feedback* mechanisms. One such instance is the output of the posterior pituitary hormone *oxytocin.* This hormone causes the muscle layer of the uterus, the *myometrium,* to contract during childbirth. Contraction of the myometrium causes additional oxytocin to be released to aid in the contraction, regardless of the amount of hormone already present.

Studying the effects of hormones on the body is difficult to do in a wet lab, since experiments can often take days, weeks, or even months to complete, and are quite expensive. In addition, live animals may need to be sacrificed, and technically difficult surgical procedures are sometimes necessary. The PhysioEx simulations you will be using in this lab will allow you to study the effects of given hormones on the body by using "virtual" animals rather than live ones. You will be able to carry out delicate surgical techniques with the click of a button. You will also be able to complete experiments in a fraction of the time that it would take in an actual wet lab environment.

Hormones and Metabolism

Metabolism is the broad term used for all biochemical reactions occurring in the body. Metabolism involves *catabolism,* a process by which complex materials are broken down into simpler substances, usually with the aid of enzymes found in body cells. Metabolism also involves *anabolism,* in which the smaller materials are built up by enzymes to build larger, more complex molecules. When bonds are broken in catabolism, energy that was stored in the bonds is released for use by the cell. When larger molecules are made, energy is stored in the various bonds formed. Some of the energy liberated may go to the formation of ATP, the energy-rich material used by the body to run itself. However, not all of the energy liberated goes into this pathway. Some of that energy is given off as body heat. Humans are *homeothermic* animals, meaning they have a fixed body temperature. Maintaining this temperature is very important to maintaining the metabolic pathways found in the body.

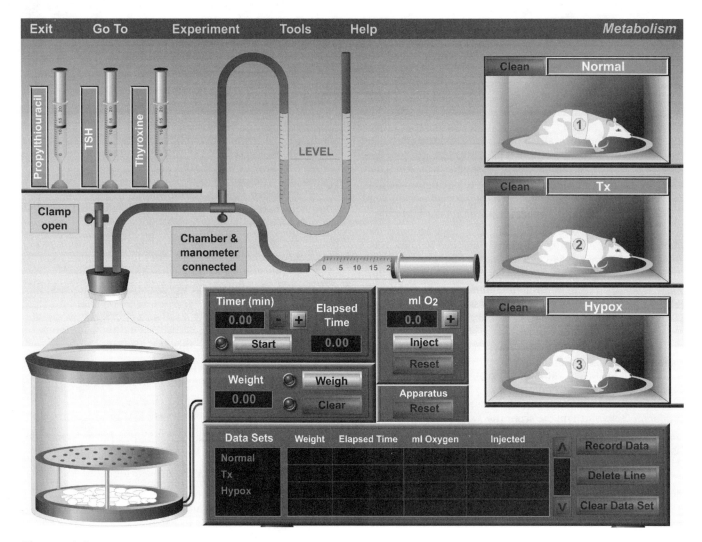

Figure 4.1   Opening screen of the Metabolism experiment.

The most important hormone in maintaining metabolism and body heat is **thyroxine.** Also known as *tetraiodothyronine,* or T_4, thyroxine is secreted by the thyroid gland, located in the neck. However, production of thyroxine is really controlled by the pituitary gland, which secretes **thyroid stimulating hormone (TSH).** TSH is carried by the blood to the thyroid gland (its *target tissue*) and causes the thyroid to produce more thyroxine.

It is also important to understand the role of the **hypothalamus** in thyroxine and TSH production. The hypothalamus, located in the brain, is a primary endocrine gland that secretes several hormones affecting the pituitary gland (also located in the brain.) Among these hormones is **thyrotropin releasing hormone (TRH),** which stimulates production of TSH in the pituitary gland. If the hypothalamus determines that there is not enough thyroxine circulating to maintain the body's metabolism, it will secrete TRH to stimulate production of TSH by the pituitary gland, which in turn will stimulate production of thyroxine by the thyroid (a classic example of a negative feedback loop). TRH travels from the hypothalamus to the pituitary gland via the **hypothalamic-pituitary portal system,** a specialized arrangement of blood vessels

consisting of a single **portal vein** that connects two capillary beds. The hypothalamic-pituitary portal system transports many other hormones from the hypothalamus to the pituitary gland. Primarily, the hormones secreted by the hypothalamus are *tropic* (or *trophic*) hormones, which are hormones that stimulate or inhibit the secretion of other hormones. TRH is an example of a tropic hormone, since it stimulates the release of TSH (which is itself a tropic hormone, since it stimulates the release of thyroxine).

In the following experiments you will be investigating the effects of thyroxine and TSH on an animal's metabolic rate. To begin, follow the instructions for starting PhysioEx in the "Getting Started" section at the front of this manual. From the main menu, select **Endocrine System Physiology.** The opening screen will appear in a few seconds (see Figure 4.1.) Select **Balloons On** from the Help menu for help identifying the equipment onscreen (you will see labels appear as you roll over each piece of equipment). Select **Balloons Off** to turn this feature off before you begin the experiments.

Study the screen. You will see a jar-shaped chamber to the left, connected to a *respirometer-manometer apparatus* (consisting of a U-shaped tube, a syringe, and associated tubing.) You will be placing animals—in this case, rats—in the chamber in order to gather information about how thyroxine and TSH affect their metabolic rates. Note that the chamber also includes a weight scale, and that next to the chamber is a timer for setting and timing the length of a given experiment. Under the timer is a weight display.

Two tubes are connected to the top of the chamber. The left tube has a clamp on it that can be opened or closed. Leaving the clamp open will allow outside air into the chamber; closing the clamp will create a closed, airtight system. The other tube leads to a *T-connector*. One branch of the T leads to a fluid-containing U-shaped tube, called a *manometer*. As an animal uses up the air in the closed system, this fluid will rise in the left side of the U-shaped tube and fall in the right.

The other branch of the T-connector leads to a syringe filled with air. Using the syringe to inject air into the tube, you will measure the amount of air that is needed to return the fluid columns to their original levels. This measurement will be equal to the amount of oxygen used by the animal during the elapsed time of the experiment. Soda lime, found at the bottom of the chamber, absorbs the carbon dioxide given off by the animal so that the amount of oxygen used can be measured easily. The amount of oxygen used by the animal, along with its weight, will be used to calculate the animal's metabolic rate.

Also on the screen are three white rats in their individual cages. These are the specimens you will use in the following experiments. One rat is **normal;** the second is **thyroidectomized** (abbreviated on the screen as **Tx**)—meaning its thyroid has been removed; and the third is **hypophysectomized** (abbreviated on the screen as **Hypox**)—meaning its pituitary gland has been removed. The pituitary gland is also known as the *hypophysis,* and removal of this organ is called a *hypophysectomy.*

To the top left of the screen are three syringes containing various chemicals: propylthiouracil, thyroid stimulating hormone (TSH), and thyroxine. TSH and thyroxine have been previously mentioned; propylthiouracil is a drug that inhibits the production of thyroxine by blocking the incorporation of iodine into the hormone. You will be performing four experiments on each animal: 1) you will determine its baseline metabolic rate, 2) you will determine its metabolic rate after it has been injected with thyroxine, 3) you will determine its metabolic rate after it has been injected with TSH, and 4) you will determine its metabolic rate after it has been injected with propylthiouracil.

You will be recording all of your data on **Chart 1** (see p. 34). You may also record your data onscreen by using the equipment in the lower part of the screen, called the *data collection unit*. This equipment records and displays the data you accumulate during the experiments. The data set for **Normal** should be highlighted in the **Data Sets** window, since you will be experimenting with the normal rat first. The **Record Data** button lets you record data after an experimental trial. Clicking the **Delete Line** or **Clear Data Set** buttons erases any data you want to delete.

Activity 1:
Determining the Baseline Metabolic Rates

First, you will determine the baseline metabolic rate for each rat.

1. Using the mouse, click and drag the **normal** rat into the chamber and place it on top of the scale. When the animal is in the chamber, release the mouse button.

2. Be sure the clamp on the left tube (on top of the chamber) is open, allowing air to enter the chamber. If the clamp is closed, click on it to open it.

3. Be sure the indicator next to the T-connector reads "Chamber and manometer connected." If not, click on the **T-connector knob.**

4. Click on the **Weigh** button in the box to the right of the chamber to weigh the rat. Record this weight in the **Baseline** section of **Chart 1** for "Weight."

5. Click the (**+**) button on the Timer so that the Timer display reads 1.00.

6. Click on the clamp to close it. This will prevent any outside air from entering the chamber, and ensure that the only oxygen the rat is breathing is the oxygen inside the closed system.

7. Click **Start** on the Timer display. You will see the elapsed time appear in the "Elapsed Time" display. Watch what happens to the water levels in the U-shaped tube.

8. At the end of the 1-minute period, the timer will automatically stop. When it stops, click on the **T-connector knob** so that the indicator reads "Manometer and syringe connected."

9. Click the clamp to open it so that the rat can once again breathe outside air.

10. Click the (**+**) button under "ml O_2" below the syringe, so that the display reads 1.0 ml. Then click **Inject,** and watch what happens to the fluid levels. Continue clicking the (**+**) button and injecting air until the fluid in the two arms of the U-tube is level again. How many ml of air needed to be added to level the fluid in the two arms? (This is equivalent to the amount of oxygen that the rat used up during the 1 minute in the closed chamber.) Record this measurement in the **Baseline** section of **Chart 1** for "ml O_2 used in 1 minute."

11. Determine the oxygen consumption per hour for the rat. Use the following formula:

$$\frac{\text{ml } O_2 \text{ consumed}}{1 \text{ minute}} \times \frac{60 \text{ minutes}}{\text{hr}} = \text{ml } O_2/\text{hr}$$

Record this data in the Baseline section of Chart 1 for "ml O2 used per hour."

12. Now that you have the amount of oxygen used per hour, determine the metabolic rate per kilogram of body weight by using the following formula (Note that you will need to convert the weight data from g to kg before you can use the formula):

$$\text{Metabolic rate} = \frac{\text{ml } O_2/\text{hr}}{\text{wt. in kg}} = \underline{\hspace{2cm}} \text{ml } O_2/\text{kg/hr}$$

Record this data in the Baseline section of Chart 1 for "Metabolic rate."

13. Click **Record Data.**

14. Click and drag the rat from the chamber back to its cage.

15. Click the **Reset** button in the box labeled *Apparatus.*

Chart 1

	Normal Rat	Thyroidectomized Rat	Hypophysectomized Rat
Baseline			
Weight	_____ grams	_____ grams	_____ grams
ml O_2 used in 1 minute	_____ ml	_____ ml	_____ ml
ml O_2 used per hour	_____ ml	_____ ml	_____ ml
Metabolic rate	_____ ml O_2 /Kg./Hr.	_____ ml O_2 /Kg./Hr.	_____ ml O_2 /Kg./Hr.
With Thyroxine			
Weight	_____ grams	_____ grams	_____ grams
ml O_2 used in 1 minute	_____ ml	_____ ml	_____ ml
ml O_2 used per hour	_____ ml	_____ ml	_____ ml
Metabolic rate	_____ ml O_2 /Kg./Hr.	_____ ml O_2 /Kg./Hr.	_____ ml O_2 /Kg./Hr.
With TSH			
Weight	_____ grams	_____ grams	_____ grams
ml O_2 used in 1 minute	_____ ml	_____ ml	_____ ml
ml O_2 used per hour	_____ ml	_____ ml	_____ ml
Metabolic rate	_____ ml O_2 /Kg./Hr.	_____ ml O_2 /Kg./Hr.	_____ ml O_2 /Kg./Hr.
With Propylthiouracil			
Weight	_____ grams	_____ grams	_____ grams
ml O_2 used in 1 minute	_____ ml	_____ ml	_____ ml
ml O_2 used per hour	_____ ml	_____ ml	_____ ml
Metabolic rate	_____ ml O_2 /Kg./Hr.	_____ ml O_2 /Kg./Hr.	_____ ml O2 /Kg./Hr.

16. Now repeat steps 1–15 for the thyroidectomized ("Tx") and hypophysectomized ("Hypox") rats. Record your data in the **Baseline** section of **Chart 1** under the corresponding column for each rat. Be sure to highlight **Tx** under **Data Sets** (on the data collection box) before beginning the experiment on the thyroidectomized rat; likewise, highlight **Hypox** under **Data Sets** before beginning the experiment on the hypophysectomized rat.

How did the metabolic rates of the three rats differ?

_____

_____

_____

Why did the metabolic rates differ?

_____

_____

_____

_____

_____

If an animal has been thryoidectomized, what hormone(s) would be missing from its blood?

_____

As a result of the missing hormone(s), what would the overall effect on the body be?

_____

_____

_____

How could you treat a thyroidectomized animal so that it functioned like a "normal" animal?

_____

_____

_____

_____

If an animal has been hypophysectomized, what effect would you expect to see in the hormone levels in its body?

_____

_____

_____

_____

What would be the effect of a hypophysectomy on the metabolism of an animal?

_____

_____

_____

_____

_____

Activity 2:
Determining the Effect of Thyroxine on Metabolic Rate

Next you will investigate the effects of **thyroxine** injections on the metabolic rates of all three rats.

Please note that in a wet lab environment you would normally need to inject thyroxine (or any other hormone) into a rat *daily* for at least 1–2 weeks in order for any response to be seen. However, in the following simulations you will only inject the rat once and will be able to witness the same results as if you had administered multiple injections over the course of several weeks. In addition, by clicking the **Clean** button while a rat is inside its cage, you can immediately remove all residue of any previously injected hormone from the rat and perform a new experiment on the same rat. In a real wet lab environment you would need to either wait weeks for hormonal residue to leave the rat's system or use a different rat.

1.   Select a rat to test. You will eventually test all three, and it doesn't matter what order you test them in. Under **Data Sets,** highlight **Normal, Tx,** or **Hypox** depending on which rat you select.

2.   Click the **Reset** button in the box labeled *Apparatus.*

3.   Click on the syringe labeled **thyroxine** and drag it over to the rat. Release the mouse button. This will cause thyroxine to be injected into the rat.

4.   Click and drag the rat back into the chamber. Perform steps 1–12 of Activity 1 again, except that this time, record your data in the **With Thyroxine** section of Chart 1.

5.   Click **Record Data.**

6.   Click and drag the rat from the chamber back to its cage, and click **Clean** to cleanse it of all traces of thyroxine.

7.   Now repeat steps 1–6 for the remaining rats. Record your data in the **With Thyroxine** section of Chart 1 under the corresponding column for each rat.

What was the effect of thyroxine on the normal rat's metabolic rate? How does it compare to the normal rat's baseline metabolic rate?

_____

_____

_____

_____

Why was this effect seen?

_____

_____

_____

_____

What was the effect of thyroxine on the thyroidectomized rat's metabolic rate? How does it compare to the thyroidectomized rat's baseline metabolic rate?

_____

_____

_____

_____

Why was this effect seen?

_____

_____

_____

_____

What was the effect of thyroxine on the hypophysectomized rat's metabolic rate? How does it compare to the hypophysectomized rat's baseline metabolic rate?

_____

_____

_____

_____

_____

Why was this effect seen?

_____

_____

_____

_____

_____ ∎

Activity 3:
Determining the Effect of TSH on Metabolic Rate

Next you will investigate the effects of TSH injections on the metabolic rates of the three rats. Select a rat to experiment on first, and then proceed.

1. Under **Data Sets,** highlight **Normal, Tx,** or **Hypox,** depending on which rat you are using.

2. Click the **Reset** button in the box labeled *Apparatus.*

3. Click and drag the syringe labeled **TSH** over to the rat and release the mouse button, injecting the rat.

4. Click and drag the rat into the chamber. Perform steps 1–12 of Activity 1 again. Record your data in the **With TSH** section of Chart 1.

5. Click **Record Data.**

6. Click and drag the rat from the chamber back to its cage, and click **Clean** to cleanse it of all traces of TSH.

7. Now repeat this activity for the remaining rats. Record your data in the **With TSH** section of Chart 1 under the corresponding column for each rat.

What was the effect of TSH on the normal rat's metabolic rate? How does it compare to the normal rat's baseline metabolic rate?

_____

_____

_____

_____

_____

_____

Why was this effect seen?

_____

_____

_____

_____

What was the effect of TSH on the thyroidectomized rat's metabolic rate? How does it compare to the thyroidectomized rat's baseline metabolic rate?

_____

_____

_____

_____

Why was this effect seen?

_____

_____

_____

What was the effect of TSH on the hypophysectomized rat's metabolic rate? How does it compare to the hypophysectomized rat's baseline metabolic rate?

_____

_____

_____

_____

Why was this effect seen?

_____

_____

_____ ■

Activity 4:
Determining the Effect of Propylthiouracil on Metabolic Rate

Next you will investigate the effects of propylthiouracil injections on the metabolic rates of the three rats. Keep in mind that propylthiouracil is a thyroxine inhibitor.

Select a rat to experiment on first, and then proceed.

1.   Under **Data Sets,** highlight **Normal, Tx,** or **Hypox,** depending on which rat you are using.

2.   Click the **Reset** button in the box labeled _Apparatus._

3.   Click and drag the syringe labeled **Propylthiouracil** over to the rat and release the mouse button, injecting the rat.

4.   Click and drag the rat into the chamber. Perform steps 1–12 of Activity 1 again, except this time record your data in the **With Propylthiouracil** section of Chart 1.

5.   Click **Record Data.**

6.   Click and drag the rat from the chamber back to its cage, and click **Clean** to cleanse it of all traces of propylthiouracil.

7.   Now repeat this activity for the remaining rats. Record your data in the **With Propylthiouracil** section of Chart 1 under the corresponding column for each rat.

8.   Click **Tools → Print Data** to print your data.

What was the effect of propylthiouracil on the normal rat's metabolic rate? How does it compare to the normal rat's baseline metabolic rate?

_____

_____

_____

Why was this effect seen?

_____

_____

_____

What was the effect of propylthiouracil on the thyroidectomized rat's metabolic rate? How does it compare to the thyroidectomized rat's baseline metabolic rate?

_____

_____

_____

Why was this effect seen?

_____

_____

_____

What was the effect of propylthiouracil on the hypophysectomized rat's metabolic rate? How does it compare to the hypophysectomized rat's baseline metabolic rate?

_____

_____

_____

Why was this effect seen?

_____

_____ ■

Hormone Replacement Therapy

Ovaries are stimulated by **follicle stimulating hormone (FSH),** released from the pituitary gland, to get ovarian follicles to develop so that they can ovulate and perhaps be fertilized. While the follicles are developing, the follicular cells that form around the oocyte produce the hormone **estrogen.** One main target tissue for estrogen is the uterus, and the action of estrogen is to enable the uterus to grow and develop so that it may receive fertilized eggs for implantation. _Ovariectomy,_ the removal of ovaries, will remove the source of estrogen and cause the uterus to slowly atrophy.

FSH and the subsequent estrogens are also regulated by a negative feedback system. Recall that hormone levels are monitored by the hypothalamus in the brain. If the hypothalamus determines that not enough FSH is present, it will release **gonadotropin releasing hormone (GnRH)** to stimulate FSH production in the pituitary gland. GnRH will be transported from the hypothalamus to the pituitary gland via the hypothalamic-pituitary portal system. Birth control pills

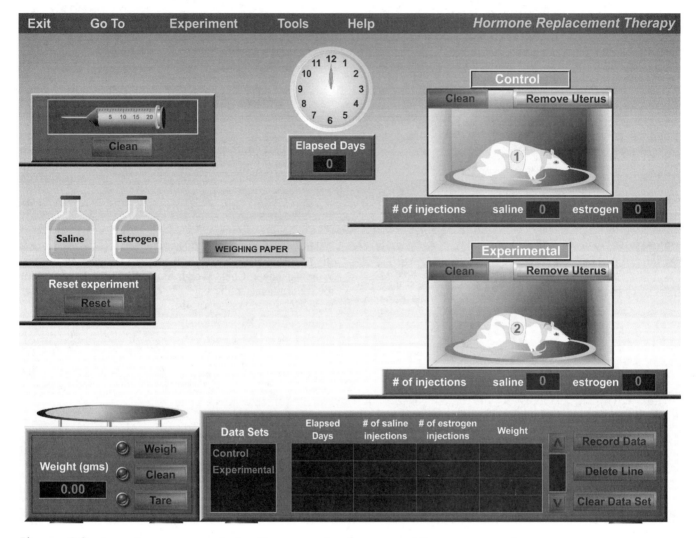

Figure 4.2 Opening screen of the Hormone Replacement Therapy experiment.

work by employing the negative feedback loop. The pills release high levels of estrogen into the body, which block the release of FSH from the pituitary gland so that no new oocyte is developed and ovulated. A similar process occurs during pregnancy: high levels of estrogen produced during pregnancy block the release of additional FSH.

In this activity you will re-create a classic endocrine experiment and examine how **estrogen** affects uterine tissue growth. You will be working with two female rats, both of which have been ovariectomized and, as a result, are no longer producing estrogen. You will administer **hormone replacement therapy** to one rat by giving it daily injections of estrogen. The other rat will serve as your "control" and receive daily injections of saline. You will then remove the uterine tissues from both rats, weigh the tissues, and compare them to determine the effects of hormone replacement therapy.

In the first ovariectomized rat, what do you think will happen to the uterus if estrogen therapy is administered?

_____

_____

_____

_____

In the second ovariectomized rat, what do you think the administration of saline will do to the uterus?

_____

_____

_____

Start by selecting **Hormone Replacement Therapy** from the **Experiment** menu. A new screen will appear (Figure 4.2), showing the two ovariectomized rats in cages. (Please note that if this were a wet lab, the ovariectomies would need to have been performed on the rats a month or more prior to the rest of the experiment in order to ensure that no residual hormones remained in the rats' systems.) Also on screen are a bottle of saline, a bottle of estrogen, a syringe, a box of weighing paper, and a weighing scale.

Proceed carefully with this experiment. Each rat will disappear from the screen once you remove its uterus, and it cannot be brought back unless you restart the experiment. This replicates the situation you would encounter if working with live animals: once the uterus is removed, the animal would have to be sacrificed.

Activity 5:
Hormone Replacement Therapy

1. Click on the syringe, drag it to the bottle of **saline,** and release the mouse button. The syringe will automatically fill with 1 ml of saline.

2. Drag the syringe to the **Control** rat and place the tip of the needle in the rat's lower abdominal area. Injections into this area are considered *interperitoneal* and will quickly be picked up by the abdominal blood vessels. Release the mouse button—the syringe will empty into the rat and automatically return to its holder. Click **Clean** on the syringe holder to "clean" the syringe of all residue.

3. Click on the syringe again, this time dragging it to the bottle of **estrogen,** and release the mouse button. The syringe will automatically fill with 1 ml of estrogen.

4. Drag the syringe to the **Experimental** rat and place the tip of the needle in the rat's lower abdominal area. Release the mouse button—the syringe will empty into the rat and automatically return to its holder. Click **Clean** on the syringe holder to "clean" the syringe of all residue.

5. Click the **Clock** above the **Elapsed Days** display You will notice the hands sweep the clock face twice, indicating that 24 hours have passed.

6. Repeat steps 1–5 until each rat has received a total of 7 injections over the course of 7 days (1 injection per day.) Note that the **# of injections** display below each rat cage records how many injections the rat has received. The control rat should receive 7 injections of saline, whereas the experimental rat should receive 7 injections of estrogen.

7. Next, click on the box of weighing paper. You will see a small piece of paper appear. Click and drag this paper over to the top of the scale and release the mouse button.

8. Notice that the scale will give you a weight for the paper. With the mouse arrow, click on the **Tare** button to tare the scales to zero (0.00 gms), adjusting for the weight of the paper.

9. You are now ready to remove the uteruses. In a wet lab, this would require surgery. Here you will simply click on the **Remove Uterus** button found in each rat cage. The rats will disappear, and a uterus (consisting of a uterine body and two uterine horns) will appear in each cage.

10. Click and drag the uterus from the Control rat over to the scale and release it on the weighing paper. Click on the **Weigh** button to obtain the weight. Record the weight here:

Uterus weight (Control): _____ gms

11. Click **Record Data.**

12. Click **Clean** on the weight scale to dispense of the weighing paper and uterus.

13. Repeat steps 7 and 8. Then click and drag the uterus from the Experimental rat over to the scale and release it on the weighing paper. Click **Weigh** to obtain the weight. Record the weight here:

Uterus weight (Experimental): _____ gms

14. Click **Record Data.**

15. Click **Clean** on the weight scale to dispense of the weighing paper and uterus.

16. Click **Tools → Print Data** to print your data.

How does the Control uterus weight compare to the Experimental uterus weight?

_____

_____

_____

What can you conclude about the administration of estrogen injections on the experimental animal?

_____

_____

_____

What might be the effect if testosterone had been administered instead of estrogen? Explain your answer.

_____

_____

_____

_____

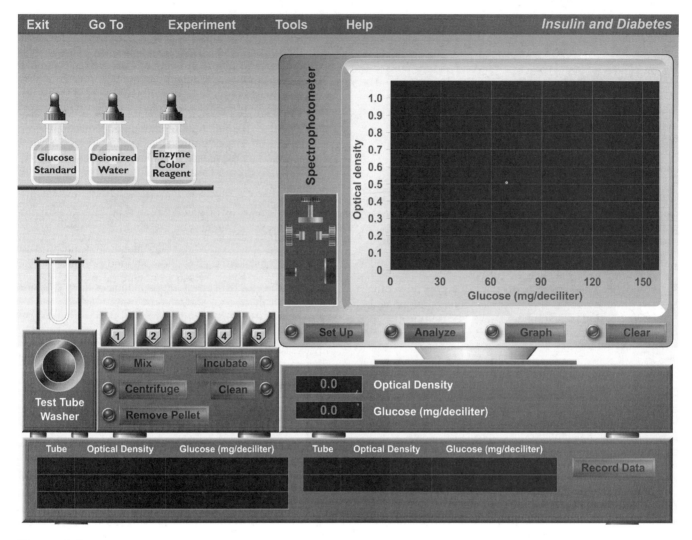

Figure 4.3   **Opening screen of the Insulin and Diabetes experiment, Part 1.**

Insulin and Diabetes

Insulin is produced by the β-cells of the endocrine portion of the pancreas. It is vital to the regulation of blood glucose levels because it enables the body's cells to absorb glucose from the bloodstream. Glucose absorbed from the blood can enter cells (usually liver or muscle cells), where excess glucose is used to form glycogen (animal starch). It is estimated that 75% of glucose taken in with a meal is stored in this manner. Since humans are considered "discontinuous feeders," this production of animal starch after a meal ensures that a glucose supply will be available for several hours after intake. The body has to maintain a certain level of glucose in the blood to serve nervous system cells, into which only glucose can be absorbed. When glucose levels in the blood fall below a certain point, the α cells of the pancreas then produce **glucagon.** The job of this hormone is to break the stored glycogen down into glucose to be released into the blood.

When insulin is not produced by the pancreas, **diabetes mellitus Type I** results. When insulin *is* produced by the pancreas but the body fails to respond to it, **diabetes mellitus**

Type II results. In either case, glucose remains in the bloodstream, unable to be taken up by the body's cells to serve as the primary fuel for metabolism. Excess glucose in the blood is then filtered by the kidney. Since the re-uptake of filtered glucose involves a finite number of receptors in kidney cells, some excess glucose will not be re-absorbed into the body and will instead pass out of the body in urine. The lack of insulin for glucose transport also affects muscle, and results in muscle cells undergoing protein catabolism so that the freed amino acids can form glucose within the liver. This action puts the body into a negative nitrogen balance from the resulting protein depletion and tissue wasting. Also associated with this condition is poor resistance to infections.

In the following experiment you will be studying the effects of insulin treatment for diabetes Type I. The experiment is divided into two parts. In Part I you will be obtaining a **glucose standard curve,** which will be explained shortly. In Part II you will compare the glucose levels of a normal rat to those of a diabetic rat, and then compare them again after each rat has been injected with insulin.

Part I

Activity 6:
Obtaining a Glucose Standard Curve

To begin, select **Insulin and Diabetes—Part 1** from the **Experiments** menu (Figure 4.3). Select **Balloons On** from the Help menu for help identifying the equipment onscreen (you will see labels appear as you roll over each piece of equipment). Select **Balloons Off** to turn this feature off before you begin the experiments.

On the right side of the opening screen is a special **spectrophotometer.** The spectrophotometer is one of the most widely used research instruments in biology. It is used to measure the amounts of light of different wavelengths absorbed and transmitted by a pigmented solution. Inside of the spectrophotometer is a source of white light, which is separated into various wavelengths (or colors) by a prism. The user selects a wavelength (color), and light of this color is passed through a tube, or *cuvette,* containing the sample being tested. (For this experiment, the spectrophotometer light source will be pre-set for a wavelength of 450 nm.) The light transmitted by the sample then passes onto a photoelectric tube, which converts the light energy into an electrical current. The current is then measured by a meter. Alternatively, the light may be measured before the sample is put into the light path, and the amount of light absorbed—called **optical density**—may then be measured. Using either method, the change in light transmittance or light absorbed can be used to measure the amount of a given substance in the sample being tested.

In Part II you will be using the spectrophotometer to determine how much glucose is present in blood samples that you will be taking from two rats. But before you can do that, you must first obtain a **glucose standard curve** so that you have a point of reference for converting optical density readings into glucose readings (which will be measured in mg/deciliter). To do this you will prepare five test tubes that contain known amounts of glucose: 30 mg/deciliter, 60 mg/deciliter, 90 mg/deciliter, 120 mg/deciliter, and 150 mg/deciliter, respectively. You will then use the spectrophotometer to determine the corresponding optical density readings for each of these known amounts of glucose. You will then use this information to perform Part II.

Also on the screen are three dropper bottles, a test tube washer, a test tube dispenser (on top of the washer), and a test tube incubation unit that you will need to prepare the samples for analysis.

1.   Click and drag the test tube (on top of the test tube washer) into slot 1 of the incubation unit. You will see another test tube pop up from the dispenser. Click and drag this second test tube into slot 2 of the incubation unit. Repeat until you have dragged a total of five test tubes into the five slots in the incubation unit.

2.   Click and hold the mouse button on the dropper cap of the **Glucose Standard** bottle. Drag the dropper cap over to tube #1. Release the mouse button to dispense the glucose. You will see that one drop of glucose solution is dropped into

the tube and that the dropper cap automatically returns to the bottle of glucose standard.

3.   Repeat step 2 with the remaining four tubes. Notice that each subsequent tube will automatically receive one additional drop of glucose standard into the tube (that is, tube #2 will receive two drops, tube #3 will receive three drops, tube #4 will receive four drops, and tube #5 will receive 5 drops).

4.   Click and hold the mouse button on the dropper cap of the **Deionized Water** bottle. Drag the dropper cap over to tube #1. Release the mouse button to dispense the water. Notice that four drops of water are automatically added to the first tube.

5.   Repeat step 4 with tubes #2, 3, and 4. Notice that each subsequent tube will receive one *less* drop of water than the previous tube (that is, tube #2 will receive three drops, tube #3 will receive two drops, and tube #4 will receive one drop). Tube #5 will receive <u>no</u> drops of water.

6.   Click on the **Mix** button of the incubator to mix the contents of the tubes.

7.   Click on the **Centrifuge** button. The tubes will descend into the incubator and be centrifuged. When tubes are centrifuged, they are spun around a center point at high speed so that any particulate matter within the tube will settle at the bottom of the tube, forming what is called a "pellet."

8.   When the tubes resurface, click on the **Remove Pellet** button. Any pellets from the centrifuging process will be removed from the test tubes.

9.   Click and hold the mouse button on the dropper cap of the **Enzyme-Color Reagent** bottle. Still holding the mouse button down, drag the dropper cap over to tube #1. When you release the mouse, you will note that five drops of reagent are added to the tube and that the stopper is returned to its bottle.

10.   Repeat step 9 for the remaining tubes.

11.   Now click **Incubate.** The tubes will descend into the incubator, where they will be shaken to completely mix the color reagent in the tube, incubate, and then resurface.

12.   Using the mouse, click on **Set Up** on the spectrophotometer. This will warm up the instrument and get it ready for your readings. In this case, "set up" also includes setting the "zero" point so the spectrophotometer will accurately read the quantity of material contained in each tube.

13.   Click and drag tube #1 into the spectrophotometer (right above the **Set Up** button) and release the mouse button. The tube will lock into place.

14.   Click **Analyze.** You will see a spot appear on the screen, and values will appear in the **Optical Density** and **Glucose** displays.

15.   Click **Record Data** on the data collection unit.

16.   Click and drag the tube into the test tube washer.

17.   Repeat steps 13–16 for the remaining test tubes.

18.   When all five tubes have been analyzed, click on the **Graph** button. This is the glucose standard graph which you will use in Part II of the experiment. ■

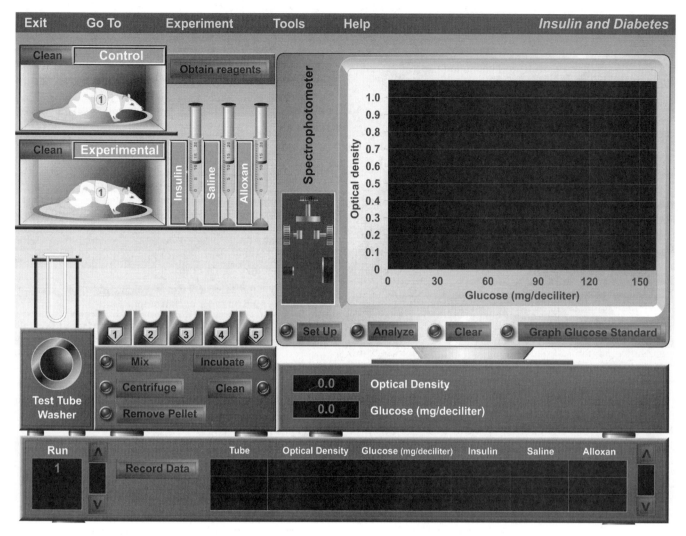

Figure 4.4 Opening screen of the Insulin and Diabetes experiment, Part 2.

Part II

Activity 7:
Comparing Glucose Levels Before and After Insulin Injection

Select **Insulin and Diabetes Part 2** from the **Experiment** menu.

The opening screen will look similar to the screen from Part I (Figure 4.4). Notice the two rats in their cages. One will be your control animal, the other will be your experimental animal. Also note the three syringes, containing insulin, saline, and alloxan, respectively. **Alloxan** is a drug that, when administered to an animal, selectively kills all the pancreatic β-cells that produce insulin, and renders the animal instantly diabetic.

In this experiment you will inject the control rat with saline and the experimental rat with alloxan. (Normally, injections are given ***daily*** for a week. In this simulation we will administer the injections only once, but we can see results as though the injections had been given over a longer period of time.)

After administering the saline and alloxan injections, you will obtain blood samples from the two rats. You will then inject both rats with insulin, and obtain blood samples again. Finally, you will analyze all the blood samples in the spectrophotometer (described in Part I) to compare the amounts of glucose present in the samples.

1. Click and drag the **Saline** syringe to the **Control** rat and release the mouse button to inject the animal.

2. Click and drag the **Alloxan** syringe to the **Experimental** rat and release the mouse button to inject the animal.

3. Click and drag a new test tube (from the test tube dispenser) over to the tail of the **Control** rat and release the mouse button. You will note three drops of blood being drawn from the tail into the tube. Next, click and drag the tube into test tube holder #1 in the incubator. (Note: The tail is a popular place to get blood from a rat. The end of the tail can easily be clipped and blood collected without really disturbing the rat. The tail heals quickly, with no harm to the animal.)

4. Click and drag another new test tube (from the test tube dispenser) over to the tail of the **Experimental** rat and release

the mouse button. Again, you will note that three drops of blood are drawn from the tail into the tube. Click and drag the tube into test tube holder **#2** in the incubator.

5. Click and drag the **Insulin** syringe to the **Control** rat and release the mouse button to inject the animal.

6. Repeat step 5 with the **Experimental** rat.

7. Repeat steps 3 and 4 again, drawing blood samples from each rat and placing the samples into test tube holders **#3** and **#4.**

8. Click the **Obtain reagents** button on the cabinet that currently displays the syringes. Both the syringes and rats will disappear, and you will see four dropper bottles in their place.

9. Click and hold the mouse button on the dropper of the **Deionized Water** bottle. Drag the dropper cap over to tube #1. Release the mouse button to dispense. You will note that five drops of water are added to the tube. This water is added so that all tubes will have the same volume.

10. Repeat step 9 for the remaining test tubes.

11. Click and hold the mouse button on the dropper of **Barium Hydroxide.** Drag the dropper cap over to tube #1. Release the mouse button to dispense. You will note that five drops of solution are added to the tube. (Barium hydroxide is used for clearing proteins and cells so that clear glucose readings may be obtained.)

12. Repeat step 11 for the remaining test tubes.

13. Click and hold the mouse button on the dropper of the **Heparin** bottle. Still holding the mouse button down, drag the dropper cap over to tube #1. Release the mouse button to dispense. Heparin is an anticoagulant that prevents the blood from clotting while being worked on.

14. Repeat step 13 for the remaining test tubes.

15. Click on the **Mix** button of the incubator to mix the contents of the tubes.

16. Click on the **Centrifuge** button. The tubes will descend into the incubator to be centrifuged and will then resurface.

17. Click on the **Remove Pellet** button to remove any pellets from the centrifuging process.

18. Click and hold the mouse button on the dropper of the **Enzyme Color Reagent** bottle. Drag the dropper cap to tube #1. Release the mouse to dispense.

19. Repeat step 18 with the remaining test tubes. In an actual wet lab, you would also shake the test tubes after adding the enzyme color reagent.

20. Click **Incubate** one more time. The tubes will descend into the incubator, incubate, and then resurface.

21. Click on **Set Up** on the spectrophotometer. This will warm up the instrument and get it ready for your readings.

22. Click **Graph Glucose Standard.** The graph from Part I of the experiment will appear on the monitor.

23. Click and drag tube #1 to the spectrophotometer and release the mouse button. The tube will lock into place.

24. Click **Analyze.** You will see a horizontal line appear on the screen and a value appear in the **Optical Density** display.

25. Drag the **moveable rule** (the red vertical line on the far right of the spectrophotometer monitor) over to where the horizontal line (from step 24) crosses the glucose standard line. Notice what happens to the **Glucose** display as you move the moveable rule to the left.

What is the glucose reading for where the horizontal line crosses the glucose standard line?

Test tube #1: _____ mg/deciliter glucose

This is your glucose reading for the sample being tested.

26. Click **Record Data** on the data collection unit.

27. Click and drag the test tube from the spectrophotometer into the test tube washer, then click **Clear** beneath the oscilloscope display.

28. Repeat steps 22–27 for the remaining test tubes. Record your glucose readings for each test tube here:

Test tube #2: _____ mg/deciliter glucose

Test tube #3: _____ mg/deciliter glucose

Test tube #4: _____ mg/deciliter glucose

How does the glucose level in test tube #1 compare to the level in test tube #2? Recall that tube #1 contains a sample from your control rat (which received injections of saline), and that tube #2 contains a sample from your experimental rat (which received injections of alloxan).

_____

_____

_____

Provide an explanation for this result:

_____

_____

What is the condition that alloxan has caused in the experimental rat?

_____

_____

How does the glucose level in test tube #3 compare to the level in test tube #1?

_____

_____

_____

Provide an explanation for this result:

_____

_____

How does the glucose level in test tube #4 compare to the level in test tube #2?

_____

_____

_____

Provide an explanation for this result:

_____

_____

_____

What was the effect of administering insulin to the control animal?

_____

_____

_____

What was the effect of administering insulin to the experimental animal?

_____

_____

_____

Click **Tools** → **Print Data** to print your recorded data. ■

Histology Review Supplement

Turn to p. 138 for a review of endocrine tissue.

Cardiovascular Dynamics

T he cardiovascular system is composed of a pump—the heart—and blood vessels that distribute blood containing oxygen and nutrients to every cell of the body. The principles governing blood flow are the same physical laws that apply to the flow of liquid through a system of pipes. For example, one very elementary law in fluid mechanics is that the flow rate of a liquid through a pipe is directly proportional to the difference between the pressures at the two ends of the pipe (the **pressure gradient**) and inversely proportional to the pipe's **resistance** (a measure of the degree to which the pipe hinders or resists the flow of the liquid):

$$\text{Flow} = \text{pressure gradient/resistance} = \Delta P/R$$

This basic law applies to blood flow as well. The "liquid" is blood, and the "pipes" are blood vessels. The pressure gradient is the difference between the pressure in arteries and that in veins that results when blood is pumped into arteries. Blood flow rate is directly proportional to the pressure gradient, and inversely proportional to resistance.

Recall that resistance is a measure of the degree to which the blood vessel hinders or resists the flow of blood. The main factors governing resistance are 1) blood vessel *radius,* 2) blood vessel *length,* and 3) blood *viscosity.*

Radius. The smaller the blood vessel radius, the greater the resistance, due to frictional drag between the blood and the vessel walls. Contraction, or *vasoconstriction,* of the blood vessel results in a decrease in the blood vessel radius. Lipid deposits can cause the radius of an artery to decrease, preventing blood from reaching the coronary tissue and result in a heart attack. Alternately, relaxation, or *vasodilation,* of the blood vessel causes an increase in the blood vessel radius. As we will see, blood vessel radius is the single most important factor in determining blood flow resistance.

Length. The longer the vessel length, the greater the resistance—again, due to friction between the blood and vessel walls. The length of a person's blood vessels change only as a person grows; otherwise, length generally remains constant.

Viscosity. Viscosity is blood "thickness," determined primarily by *hematocrit*—the fractional contribution of red blood cells to total blood volume. The higher the hematocrit, the greater the viscosity. Under most physiological conditions, hematocrit does not vary by much, and blood viscosity remains more or less constant.

A fourth factor in resistance is the manner of blood flow. In *laminar* flow, blood flows calmly and smoothly along the length of the vessel. In *turbulent* flow, blood flows quickly and roughly. Most blood flow in the body is laminar, and the experiments we will conduct in this lab focus on laminar flow.

Objectives

1. To understand the relationships among **blood flow, pressure gradient,** and **resistance**
2. To define resistance and describe the main factors affecting resistance
3. To describe Poiseuille's equation and how it relates to cardiovascular dynamics
4. To define **diastole, systole, end systolic volume, end diastolic volume, stroke volume, isovolumetric contraction,** and **ventricular ejection**
5. To describe **Starling's Law** and its application to cardiovascular dynamics
6. To design your own experiments using the lab simulation for pump mechanics
7. To understand what is meant by the term **compensation**

Poiseuille's equation expresses the relationships among blood pressure, vessel radius, vessel length, and blood viscosity on laminar blood flow:

$$\text{Blood flow } (\Delta Q) = \pi\Delta Pr^4/8\eta l$$

or

$$\text{Blood flow } (\Delta Q) = \frac{\pi\Delta Pr^4}{8\eta l}$$

where

ΔP = pressure difference between the two ends of the vessel
r = radius of the blood vessel
η = viscosity
l = vessel length

This equation states that changes in pressure, blood vessel radius, viscosity, and vessel length all have an effect on blood flow. Note that the effect of radius (r) on blood flow is especially strong (fluid flow varies with radius to the fourth degree).

The main method of controlling blood flow is via contraction or relaxation of the smooth muscle found in the tunica media of an artery. When contracted, the radius of the artery becomes much smaller, resulting in more resistance to blood flow within the artery. The smaller arteries and arterioles that regulate blood flow throughout the body are referred to as *resistance vessels* and are very important to maintaining arterial blood pressure. If all blood vessels were to completely relax, blood pressure would fall to very dangerous levels. It is also known that the lining of arteries, the *endothelium,* releases nitric oxide in response to rapid flow in the vessel. The nitric oxide causes a dilation of the artery that reduces the shear stress.

In our first experiment, we will take a closer look at how pressure, vessel radius, blood viscosity, and vessel length affect blood flow.

Vessel Resistance

Follow the instructions in the "Getting Started" section at the front of this manual for starting PhysioEx 3.0. From the main menu, select the fifth lab, **Cardiovascular Dynamics.** The opening screen for the "Vessel Resistance" activity will appear (see Figure 5.1).

Notice the two glass beakers and the tube connecting them. Imagine that the left beaker is your heart, the tube is an artery, and the right beaker is a destination in your body, such as another organ. Clicking the **Start** button underneath the left beaker will cause blood to begin flowing from the left beaker to the right beaker. You may adjust the radius of the tube, the viscosity of the blood, or the length of the tube by adjusting the (+) and (−) buttons next to the corresponding displays.

You may also adjust the pressure by clicking the (+) and (−) buttons for pressure on top of the left beaker. Clicking **Refill** will empty the right beaker and refill the left beaker.

At the bottom of the screen is a data recording box. Clicking **Record Data** after an experimental run will record that run's data in the box.

The Effect of Pressure on Blood Flow

$$\text{Blood flow } (\Delta Q) = \frac{\pi \Delta P r^4}{8\eta l}$$

Recall that DP in Poiseuille's equation stands for the difference in pressure between the two ends of a vessel, or the pressure gradient. In order to study the pressure gradient, you will observe how blood flows at a given pressure, then change the pressure to observe the effects of the change on the blood flow.

Figure 5.1 Opening screen of the Vessel Resistance experiment.

1.   Set the **Pressure** to 25 mm Hg by clicking the (−) button on top of the left beaker.

2.   Set the **Radius** of the tube to 5.0 mm.

3.   Set the blood **Viscosity** to 3.5.

4.   Set the vessel **Length** to 50 mm.

5.   Highlight the **Pressure** data set by clicking the word Pressure in the box at the bottom left of your screen.

6.   Make sure the left beaker is filled with blood. If not, click **Refill.**

7.   Click **Start.**

8.   When the right beaker is full, click **Record Data.** Your data will appear in the data recording box.

9.   Click **Refill.**

10.   Increase the **Pressure** by 25 mm Hg (that is, set it to 50 mm Hg.) Leave the radius, viscosity, and length settings the same. Click **Start** again, and click **Record Data** once the right beaker is full. Click **Refill.**

11.   Continue repeating the experiment, increasing the **Pressure** by 25 mm Hg each time, until you have reached 225 mm Hg. Remember to click **Record Data** after each run.

12.   Now click **Tools** at the top of the screen. A drop-down menu will appear. Highlight **Plot Data** and click it. You will see your data appear in a data plot. Note that there are two slide bars: one for the *X*-axis and one for the *Y*-axis. Set the slide bar for the *X*-axis to "Pressure" and the slide bar for the *Y*-axis to "Flow." You may wish to print the data plot by clicking **Print Plot** at the top left of the plot data window. Click the "X" at the top right of the plot data screen to close the window.

Describe the relationship between pressure and blood flow.

_____

_____

What kind of change in the cardiovascular system would result in a pressure change?

_____

_____

Why would such a change cause problems?

_____

_____

_____

_____ ■

The Effect of Vessel Radius on Blood Flow

$$\text{Blood flow } (\Delta Q) \ = \ \frac{\pi \Delta P r^4}{8 \eta l}$$

The next parameter of Poiseuille's equation we examine is vessel radius. In the equation, this parameter is taken to the fourth power (r^4). This means that a small change in vessel radius can result in a large alteration in blood flow.

1.   At the lower left corner of the screen, under "Data Sets," click **Radius.**

2.   Be sure that the left beaker is full. If not, click **Refill.**

3.   Set the **Pressure** to 100 mm Hg.

4.   Set the vessel **Length** to 50 mm.

5.   Set the blood **Viscosity** to 1.0.

6.   Set the **Radius** of the tube to 1.5 mm.

7.   Click **Start** and allow the blood to travel from the left beaker to the right beaker. When the blood has completely transferred to the right beaker, click **Record Data.**

8.   Increase the radius of the tube by 1.0 mm (set it to 2.5 mm) and repeat the experiment. Leave all the other settings the same. Continue repeating the experiment until you reach the maximum radius setting of 6.0 mm. Be sure to click **Refill** and **Record Data** after each run.

9.   Click **Tools** on top of the screen and select **Plot Data.** Again, you will see your data appear on a data plot. Slide the *X*-axis bar to the "Radius" setting and the *Y*-axis bar to the "Flow" setting. You may wish to print the data plot by clicking **Print Plot** at the top left corner of the window. Close the window by clicking the "X" on the top right corner of the window.

Describe the relationship between radius and blood flow.

_____

_____

How does this graph differ from your first graph?

_____

_____

_____

_____

In this activity, we mechanically altered the radius of the tube by clicking the (+) button next to **Radius.** Physiologically, what could cause the radius of a blood vessel to change in our bodies?

_____

_____

_____

_____

_____

In a clogged artery, what has happened to the radius of the artery? How has this affected blood flow? What could be done to fix this condition?

_____

_____

_____

_____

_____

When a blood vessel bifurcates (splits) into two smaller vessels, the radii of the two smaller vessels add up to a larger cumulative radius than the radius of the original vessel. However, blood flow is slower in the two vessels than in the original. Why?

_____

_____

_____

What is the advantage of having slower blood flow in some areas of the body, such as in the capillaries of our fingers?

_____

_____

_____

_____

A c t i v i t y   3 :
The Effect of Viscosity on Blood Flow

$$\text{Blood flow } (\Delta Q) = \frac{\pi \Delta P r^4}{8 \eta l}$$

Blood consists of *plasma* (the fluid portion of blood, containing proteins, nutrients, and other solutes) and *formed elements* (including red and white blood cells and platelets). Viscosity is a measurement of the "thickness" of blood. Plasma has a viscosity of 1.2 to 1.3 times that of water. Whole blood has about twice the viscosity of plasma alone. Blood viscosity depends mainly on *hematocrit,* or the fractional contribution of red blood cells to total blood volume. The higher the hematocrit, the more viscous the blood is; the lower the hematocrit, the less viscous the blood. In severe *anemia,* a condition characterized by a low number of red blood cells, blood viscosity is low. In *polycythemia vera,* a condition in which the number of red blood cells increases, there are abnormally high hematocrit counts, resulting in blood that can be twice as viscous as normal.

1. Set the **Pressure** to 100 mg Hg.

2. Set the vessel **Radius** to 5.00 mm.

3. Set the vessel **Length** to 50 mm.

4. Set the blood **Viscosity** to 1.0.

5. Highight **Viscosity** under "Data Sets" at the bottom left corner of the screen.

6. Make sure that the left beaker is filled with blood. If not, click **Refill.**

7. Click **Start.** After the blood has completely transferred to the right beaker, click **Record Data** and then **Refil!.**

8. Increase the **Viscosity** value by 1.0 and repeat the experiment. Leave all the other settings the same. Continue repeating the experiment until you have reached the maximum viscosity value of 10.0. Remember to click **Record Data** after each run and **Refill.**

9. Click **Tools** at the top of your screen, and select **Plot Data.**

10. Move the *X*-axis slide bar to the "Viscosity" setting, and the *Y*-axis slide bar to the "Flow" setting. You may wish to print your graph by clicking **Print Plot** at the top left corner of the window. Click the "X" at the top right corner to close the window.

Describe the relationship between viscosity and blood flow.

_____

_____

_____

_____

How does this graph compare to the previous graphs for 1) blood flow and pressure, and 2) blood flow and radius?

_____

_____

_____

_____

Predict the effect of polycythemia vera on blood flow rate.

_____

_____

How would blood viscosity alter with dehydration of the body?

_____

_____

_____

What would happen to blood flow if the body were dehydrated?

_____

_____

_____

_____

_____ ∎

Activity 4:
The Effect of Vessel Length on Blood Flow

$$\text{Blood flow } (\Delta Q) = \frac{\pi \Delta P r^4}{8 \eta l}$$

In this activity we will examine how vessel length (l) affects blood flow. Recall that longer vessels have greater resistance than shorter vessels. Shear forces between the laminar layers increase resistance and lessen flow. In humans, blood vessels change length when the body grows, but otherwise length stays constant.

1. Set the **Pressure** to 100 mm Hg.
2. Set the vessel **Radius** to 5.0.
3. Set the blood **Viscosity** to 3.5.
4. Set the vessel **Length** to 10 mm.

5. Highlight the **Length** data set at the bottom left. Be sure the left beaker is filled with blood; if not, click **Refill.**

6. Click **Start** and allow the fluid to completely transfer from the left beaker to the right beaker. When finished, click on **Record Data.** Then click **Refill.**

7. Increase the vessel **Length** by 10 mm (to 20 mm) and repeat the experiment. Continue to repeat the experiment until you have reached the maximum vessel length of 50 mm. Remember to click **Record Data** and **Refill** after each run.

8. Click **Tools** on top of the screen, then select **Plot Data.** Move the X-axis slide bar to "Length" and the Y-axis slide bar to "Flow." If you wish, click **Print Plot** on the top left of the window to print your graph. Click "X" at the top right of the window to close the window.

Describe the relationship between vessel length and blood flow.

_____

_____

_____

Why is vessel radius a more important factor in controlling blood flow resistance than vessel length?

_____

_____

_____

_____

To print the data from the data recording box, click **Tools** on top of the screen and then select **Print Data.** ∎

Pump Mechanics

The heart is an intermittent pump. The right side of the heart pumps blood to the lungs so that blood can take up oxygen. This blood is then returned to the heart, pumped by the left side of the heart to the rest of the body, and then returned to the right side of the heart again. This *cardiac cycle* happens in one heartbeat, and involves both contraction and relaxation. Blood moves into the right atrium of the heart from the vena cavae (*superior vena cava* from the head and *inferior vena cava* from the rest of the body). On the left side of the heart, blood is returned to the heart by the pulmonary veins coming from the lungs. During **diastole** (when the ventricles are relaxed), the blood entering the atria flow through the atrioventricular valves into the ventricles. The blood volume in the ventricles at the end of diastole is referred to as the **end diastolic volume** (EDV). When the atria begin to contract, the ventricles begin to contract, or enter **systole.**

At the start of systole, the pressure within the ventricles rises, due to the force of the myocardial (*myocardium* is cardiac muscle mass) walls on the enclosed blood. This rise in pressure closes the atrioventricular valves, yet is not enough to force the semilunar valves (going to the pulmonary trunk and aorta) to open. Because the blood is contained within the ventricles and the volume of blood remains constant, this is termed **isovolumetric contraction.** This ends as the pressure within the ventricles rises, so that the semilunar valves are forced open to allow the blood to leave. During the remainder of systole, the blood is forced into the pulmonary trunk and aorta, and ventricular volume falls. This ejection of blood is termed **ventricular ejection,** during which ventricular pressure rises and then begins to decline. When ventricular pressure falls below aortic pressure, the semilunar valves close, ending systole. At the end of ejection, a volume of blood approximately equal to that ejected during systole remains. This volume is referred to as the **end systolic volume** (ESV). This residual volume is fairly constant unless heart rate has increased or vessel resistance has fallen.

Cardiac output is the amount of blood each ventricle pumps per minute. During exercise, tissues need more oxygen and send neural signals to the heart to increase the heart rate. During this same exercise, respiration increases so that there is plenty of oxygen to oxygenate the blood. The changes in the thoracic cavity caused by increased respiration (especially inspiration) cause an increase in the blood returned to the heart. **Starling's Law** states that when the rate at which blood returns to the heart changes, the heart will adjust its output to match the change in inflow. As more blood is returned to the heart, the amount of blood pumped to the body per contraction per ventricle **(stroke volume)** increases. Thus, exercise leads to an increase in the stroke volume of the heart. By definition, cardiac output is the stroke volume times the number of heart beats per minute. A "normal" stroke volume is 70 ml and, with a heart rate of 75 beats per minute, the cardiac output is a little over 5 liters per minute. This is the approximate volume of blood in the body. The heart pumps this entire volume of blood in the body each minute of life.

Select **Pump Mechanics** from the **Experiment** menu at the top of the screen. You will soon see the screen shown in Figure 5.2. There are now three beakers on screen. Imagine that the left-most beaker represents blood coming from the lungs; the middle beaker represents the left side of your heart

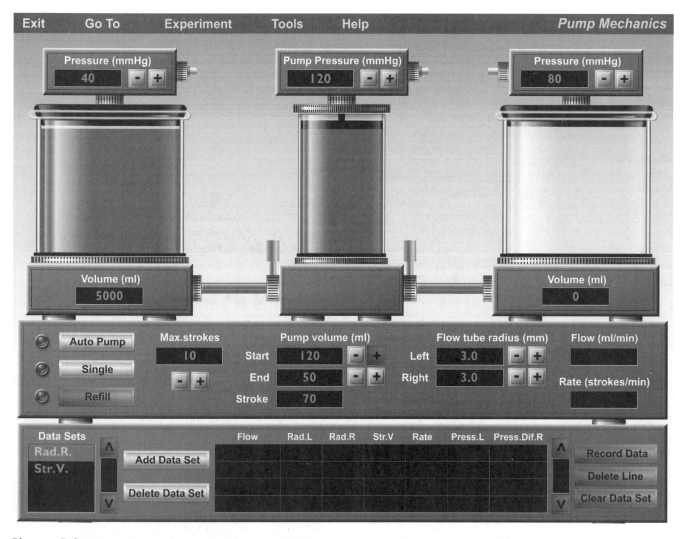

Figure 5.2 Opening screen of the Pump Mechanics experiment.

(simplified here as a single pump); and the right-most beaker represents the rest of your body, to where blood must be delivered. In between the first two beakers is a flow tube (or vessel), analogous to a vein. In between the second and third beakers is another flow tube (or vessel), analogous to an artery. One-way valves in the flow tubes ensure that blood will flow in only one direction (left to right), and these valves have flags indicating whether or not they are open or closed.

Pressure can be adjusted for each of the three beakers. The pump is governed by a pressure unit located on top of the middle beaker, which applies pressure only during the downward stroke of the pump. Upward strokes of the pump are driven by pressure from the left beaker. The pump has no resistance to flow from the left beaker. In contrast, pressure in the right beaker works against the pump pressure. Thus, the net pressure driving the fluid into the right beaker is automatically calculated by subtracting the right beaker pressure from the pump pressure. The resulting pressure difference is shown in the data recording box at the bottom of the screen, in the column labeled **Press.Dif.R.**

In this experiment you can vary the starting and ending volumes of the pump (analogous to EDV and ESV, respectively), the driving and resistance pressures (the heart pressure and blood vessel resistance), and the radii of the flow tubes leading to and from the pump chamber. Keep in mind what you learned in the earlier activity about the relationships among flow, radius, and pressure.

Clicking the **Auto Pump** button underneath the left-most beaker will cycle the pump through the number of strokes indicated in the **Max.strokes** window. Clicking the **Single** button will cycle the pump through one stroke. When performing the experiment, the pump and flow rates are automatically displayed when the number of strokes is 5 or greater. The stroke volume (**Str.V.**) of the pump is automatically computed by subtracting its ending volume from the starting volume. The ending and starting volumes can be adjusted, so the stroke volume may be adjusted by clicking on the appropriate plus or minus buttons next to **Start** and **End** under **Pump volume.**

The data recording box will record and display data accumulated during the experiments. The data for the first experiment (**Rad.R.,** which represents the right flow tube radius) should be highlighted in the "Data Sets" window. The **Record Data** button on the right edge of the screen automatically activates after an experimental trial. When clicked after a run, the **Record Data** button will display the flow rate data and save it in the computer's memory. By clicking on **Delete Line** or **Clear Data Set,** any data you want to discard may be removed from memory.

Activity 5:
Effect of Vessel Radius on Pump

In this activity, only the radius of the right flow tube leaving the pump will be manipulated. Recall that vessels leaving the heart are arteries, which have a layer of smooth muscle in the tunica media. This tunica media is stimulated by the autonomic nervous system so that the radius of the vessel will be altered depending upon the needs of the body at that particular time.

1. Make sure that **Rad. R.** is highlighted under **Data Sets** in the lower left of the screen.

2. If the left-most and middle beakers are not filled with blood, click **Refill.**

3. Set the **right Flow tube radius** to 3.0 mm.

4. Set the **left Flow tube radius** to 3.5.

5. Set the left beaker **Pressure** to 40 mm Hg.

6. Set the middle beaker **Pump Pressure** to 120 mm Hg.

7. Set the right beaker **Pressure** to 80 mm Hg.

8. Set the **Starting Pump Volume** (EDV) to 120 ml.

9. Set the **Ending Pump Volume** (ESV) to 50 ml.

10. Set **Max.strokes** to 10.

11. Click on the **Single** button and watch the pump action. Notice whether the valves are open or closed.

12. Click on **Auto Pump.** After 10 strokes have been delivered to the right beaker, the flow and rate windows will automatically display the experimental results.

13. Click **Record Data.**

14. Click **Refill.**

15. Increase the **right Flow tube radius** by 0.5 mm and repeat the experiment. Leave all the other settings the same. Continue repeating the experiment until you have reached the maximum radius of 6.0 mm. Remember to click **Record Data** and **Refill** after each run.

16. Click **Tools** at the top of the page, then **Plot Data.** Slide the X-axis bar to **Rad.R.** and the Y-axis bar to **Flow.** To print the data graph, click **Print Plot** at the top left of the window. To close the window, click the "X" at the top right corner of the window.

How does this radius plot compare to the radius plot you saw in the earlier **Vessel Resistance** activity?

_____

_____

What is the position of the pump piston during diastole?

_____

_____

What is the position of the pump piston during systole?

_____

_____

If the pump represents the left side of the heart, what does the right-most beaker represent?

_____

_____

Describe the relationship between right flow tube radius and flow.

_____

_____

How would a decrease in left flow tube radius affect flow and pump rate? Predict the outcome here.

_____

_____

_____ ■

Activity 6:
Effect of Stroke Volume on Pump

In a normal individual, 60% of the blood contained within the heart is ejected from the heart during systole, leaving 40% of the blood behind. The blood ejected by the heart is called the **stroke volume** and is the difference between the EDV and ESV (stroke volume = EDV − ESV.) Starling's Law tells us that when more blood than normal is returned to the heart by the venous system, the heart muscle will be stretched, resulting in a more forceful contraction. This, in turn, will cause more blood than normal to be ejected by the heart, raising stroke volume. In the next activity you will examine how activity of the pump is affected by changing the starting and ending volumes (and thus the stroke volume).

1. Highlight **Str.V.** under **Data Sets** in the lower left corner of the screen.

2. If the left-most beaker and middle beaker are not filled, click **Refill.**

3. Adjust the stroke volume to 10 ml by setting the starting **Pump volume** (EDV) to 120 ml and the ending **Pump volume** (ESV) to 110 ml.

4. Set the **Pressure** of the left-most beaker to 40 mm Hg.

5. Set the **Pressure** of the middle beaker (the pump) to 120 mm Hg.

6. Set the **Pressure** of the right-most beaker to 80 mm Hg.

7. Set the **Flow tube radius** to 3.0 for both the left and right flow tubes.

8. Set **Max.strokes** to 10.

9. Click **Auto Pump.** After 10 strokes have been delivered to the right beaker, the **Flow** and **Rate** windows will display the experimental results. Click **Record Data,** then click **Refill.**

10. Increase the stroke volume by increments of 10 ml by decreasing the ending **Pump volume** and repeat the experiment. Leave all the other settings the same. Continue repeating the experiment until you have reached the maximum stroke volume (120 ml). Be sure to click **Record Data** and **Refill** after each run. Watch the pump action during each stroke to see how concepts of EDV and ESV may be applied to this procedure.

11. Select **Tools** from the top of the screen, then **Plot Data.**

12. Slide the _X_-axis bar to **Str.V.** and the _Y_-axis slide to **Flow.** You may print the data plot by clicking **Print Plot** at the top left corner of the window. To close the window, click the "X" at the top right of the window.

As the stroke volume was increased, what happened to the rate of the pump?

_____

_____

_____

What would happen to the pump rate if you decreased the stroke volume?

_____

_____

_____

What would occur if blood were returned to the left side of the heart at a faster rate than it left the right side of the heart?

_____

_____

_____

_____

What might occur if the valves became constricted?

_____

_____

_____

_____ ■

Activity 7:
Compensation

If a blood vessel is compromised in some way (for example, if the vessel radius is decreased due to lipid deposits), there are ways your cardiovascular system can "compensate" for this deficiency to some degree. In this activity you will be using your knowledge of how various factors affect blood flow in order to come up with examples of compensation.

Click the **Add Data Set** button near the bottom of the screen. A small window will appear, asking you to name your new data set. Since we will be studying compensation in this activity, type in Comp as the name of your new data set. You will see this new data set appear in the **Data Sets** window. Click it to highlight it before beginning this activity. When you click **Record Data** later on, your data will be recorded within this new data set.

1. Set the **Pressure** for the left-most beaker at 40 mm Hg.

2. Set the **Pressure** for the middle beaker (the pump) at 120 mm Hg.

3. Set the **Pressure** for the right-most beaker to 80 mm Hg.

4. Set the vessel **Radius** at 3.0 mm for both the right and left flow tubes.

5. Set the **Max.strokes** at 10.

6. Set the starting **Pump volume** at 120 ml.

7. Set the ending **Pump volume** at 50 ml and click **Refill.**

8. Click **Auto Pump.** At the end of the run, click **Record Data.** This will be your "baseline" data, which you will use to compare against all subsequent experimental data.

9. Decrease the right **Flow tube radius** to 2.0 mm and repeat the experiment. Remember to click **Refill.**

How does the current flow rate compare with the baseline data?

_____

_____

Without changing the right **Flow tube radius,** what could you do to make the current flow rate equal to the flow rate from the baseline data? List three possible solutions, then test each of these on screen.

_____

_____

_____

_____

Which of your three proposed solutions was most effective?

_____

_____

In people with a high-fat diet, arteriosclerosis (a decrease in vessel diameter) is a common problem. What would the heart have to do to ensure that all organs are getting the adequate blood supply?

_____

_____

_____

_____ ■

Activity 8:
More Practice Designing Your Own Experiments

You should now have a fairly good idea of how pump mechanics work. In this section you will set up your own experimental conditions to answer a set of questions (see below). Think carefully about the structured work you have done so far. Read each question and decide how to set up the various experimental parameters in order to arrive at an answer. Then conduct an experimental run by clicking **Auto Pump** and recording your data.

You will need to create a new data set to record your data for these experiments. Click the **Add Data Set** button at the bottom of the screen. A small window will appear, asking you to name your new data set. Enter any name you like—for example, **Data Set 4.** Your newly created data set will appear in the "Data Sets" window. Highlight the name of your data set before beginning your experiments. After each experimental run, click **Record Data.** Your data will be saved in the new data set that you just created.

1. Compare the effect on flow rate of decreasing the right flow tube radius vs. the effect of decreasing the left flow tube radius (while keeping all other variables constant).

_____

_____

Recall that the flow tube between the left and middle beakers represents a vein, while the flow tube between the middle and right beakers represents an artery. In a living system, would you expect the vein or the artery to be more susceptible to a change in radius? Why?

_____

_____

_____

_____

2.   What happens to flow rate when you decrease the pressure in the left beaker?

_____

_____

Why?

_____

_____

_____

What might be a cause of pressure decrease in the left beaker?

_____

_____

3.   What happens to flow rate when you decrease the pressure in the right beaker?

_____

Why?

_____

_____

_____

What might be a cause of pressure decrease in the right beaker?

_____

_____

_____

4.   What happens to flow rate when you *increase* the pressure in the right beaker?

_____

Why?

_____

_____

_____

What might be a cause of pressure decrease in the right beaker?

_____

_____

_____

You may print your recorded data at any time by clicking **Tools** at the top of the screen, and then selecting **Print Data**. ■

Histology Review Supplement

Turn to p. 140 for a review of cardiovascular tissue.

Cardiovascular Physiology

Objectives

1. To define **autorhythmicity, sinoatrial node, pacemaker cells,** and **vagus nerves**
2. To understand the effects of the sympathetic and parasympathetic nervous systems on heart rate
3. To understand the five phases of a cardiac action potential
4. To list at least two key differences between cardiac muscle and skeletal muscle
5. To define **effective refractory period** and **relative refractory period**
6. To understand and explain the effects of each of the following on the heart: epinephrine, pilocarpine, atropine, digitalis, temperature, Na^+, Ca^{2+}, K^+

Cardiac muscle is different from other types of muscle in that cardiac muscle contracts spontaneously, without any external stimuli, whereas other muscles require signals from the nervous system in order to contract. The heart's ability to trigger its own contractions is called **autorhythmicity.** If cardiac muscle cells are isolated, placed into cell culture, and observed under the microscope, the cells can be seen to undergo contractions. No other muscle cells will contract like this unless stimulated in some manner.

Each cardiac cell has its own intrinsic rate of contraction. However, if two cells of different contraction rates are joined together, the two will contract at the rate of the faster cell. In the heart, the cells with the fastest contraction rates join together to form the **sinoatrial node (SA node)**—the "pacemaker" of the heart. Cells in the SA node, called **pacemaker cells,** direct the contraction rate of the entire heart by generating action potentials on a regular basis. These cells (and consequently, heart rate) are under the influence of the *sympathetic nervous system,* which releases the neurotransmitter norepinephrine, and the *parasympathetic nervous system,* which releases the neurotransmitter acetylcholine. Sympathetic neurons cause pacemaker cells to generate action potentials more frequently, and thus increase heart rate. In contrast, parasympathetic neurons have the opposite effect—they cause a decrease in the frequency of action potentials generated by pacemaker cells, and thus decrease heart rate. The heart simultaneously receives both sympathetic and parasympathetic signals in a "push-pull" manner: Increases in one are accompanied by decreases in the other. In healthy

adults, resting heart rate is about 70 beats per minute, although this varies depending on factors such as a person's age, physical fitness, and emotional state.

The heart receives sympathetic input via several nerves extending from the spinal cord, and receives parasympathetic input via a single pair of cranial nerves called **vagus nerves.** Once stimulated by a nerve, cardiac muscle cells depolarize. There are five main phases of depolarization (see Figure 6.1). *Phase 0* is characterized by a rapid upswing in membrane potential. Depolarization causes voltage-gated sodium channels in the cell membrane to open, increasing the flow of sodium ions into the cell and increasing the membrane potential. In *phase 1,* the open sodium channels begin to inactivate, decreasing the flow of sodium ions into the cell and causing the membrane potential to fall slightly. At the same time, however, voltage-gated potassium channels close while voltage-gated calcium channels open. The subsequent decrease in the flow of potassium out of the cell and increase in the flow of calcium into the cell act to depolarize the membrane and curb the fall in membrane potential caused by the inactivation of sodium channels. In *phase 2,* known as the *plateau phase,* the membrane remains in a depolarized state. Potassium channels stay closed and calcium channels stay open. This plateau lasts about 0.2 second. The internal potential then gradually falls to more negative values during *phase 3,* when a second set of potassium channels that began opening in phases 1 and 2 allow significant amounts of potassium to flow out of the cell. The fall in potential causes calcium channels to begin closing, reducing the flow of calcium into the cell and repolarizing the membrane until the resting potential is reached. The period of resting potential until the next depolarization is known as *phase 4.*

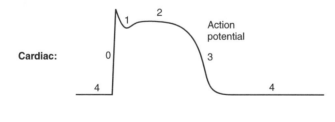

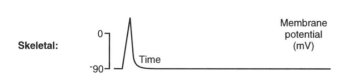

Figure 6.1 The stages of the cardiac action potential are shown on the top graph. Notice the difference between the cardiac action potential graph and the skeletal muscle action potential graph below it.

Electrical Stimulation

In the following experiments we will be examining the effects of direct stimulation and vagus nerve stimulation of a frog heart. The frog heart is a three-chambered structure, with two atria and a single ventricle. When a frog heart is removed from the frog's body, it will continue to beat for some time—convenient for experimental purposes. Although there are obvious differences between a frog heart and a human heart (e.g., human heart rate is faster), their underlying physiological mechanisms are very similar.

Follow the instructions in the "Getting Started" section at the front of this manual for starting PhysioEx 3.0. From the main menu, select the sixth lab, **Frog Cardiovascular Physiology.** The opening screen for the "Electrical Stimulation" experiment will appear (see Figure 6.2).

At the left you will see a heart suspended by a thread with a hook through the heart apex. The heartbeat is visible on the oscilloscope monitor in the upper right quarter of the screen. (Note that the heartbeat trace is *not* the same graph as the graph in Figure 6.1. The heartbeat trace shows muscle contraction, whereas Figure 6.1 shows cardiac action potential.) Observe the trace of the normal heartbeat. There is a small rise just prior to the larger rise. The smaller rise is *atrial contraction,* while the larger rise is *ventricular contraction* (Figure 6.3a). In response to stimuli being administered to the heart, the trace may exhibit *extrasystoles,* or extra beats, followed by a *compensatory pause* (Figure 6.3b).

In the lower left corner of the monitor is a **Heart Rate** window, which displays the heart rate in beats per minute (bpm). To the right of this is a status window displaying the status of the heartbeat. The status window will display one of the following: *Heart Rate Normal* (heart rate at resting condition), *Heart Rate Changing* (which will correspond to changes in the **Heart Rate** window), or *Heart Rate Stable* (heart rate is steady but is either higher or lower than at resting condition).

The white tube visible at the bottom right of the heart represents the vagus nerve. Recall that this is the nerve that delivers signals from the parasympathetic nervous system to the heart. Below the heart is a tray, and below the tray are two electrodes: one is labeled **Direct Heart Stimulation,** the other **Vagus Nerve Stimulation.** You will be clicking and

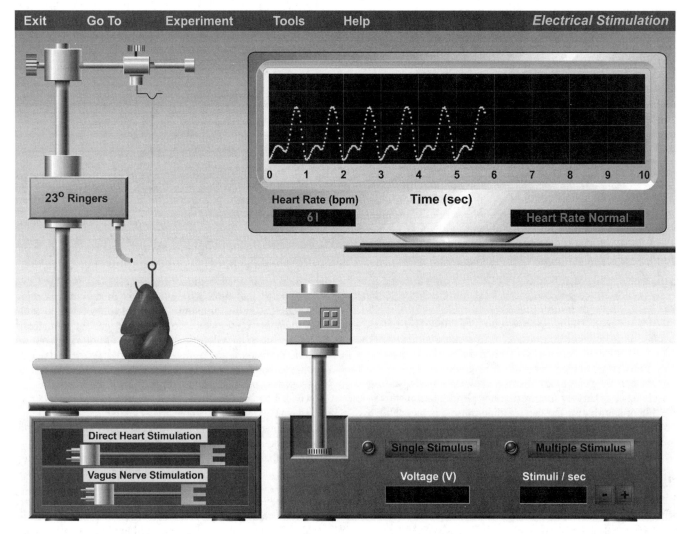

Figure 6.2 Opening screen of the Electrical Stimulation experiment.

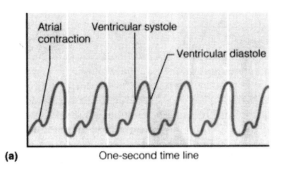

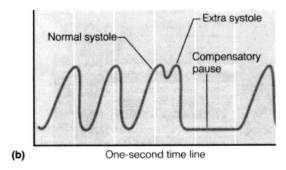

Figure 6.3 Recording of contractile activity of a frog heart. (a) Normal heartbeat. (b) Induction of an extrasystole.

dragging each of these electrodes to the electrode holder (the vertical post with four green squares). The electrode holder is emerging from a box called the *stimulator.* By clicking the **Single Stimulus** or **Multiple Stimulus** buttons after placing an electrode in the electrode holder, you will administer electrical stimuli to the frog heart.

Examine the heartbeat tracing again. This is the "baseline" heartbeat against which all others will be compared. Note that the resting heartbeat is between 59 and 61 bpm.

Activity 1:
Direct Heart Stimulation

In Exercise 2, Skeletal Muscle Physiology, we witnessed the phenomenon of *summation* in skeletal muscles. Recall that summation was the result of the skeletal muscle being stimulated with such frequency that twitches overlapped and resulted in a stronger contraction than a stand-alone twitch. This was possible because skeletal muscle has a relatively short **refractory period** (a period during which no action potentials can be generated). Unlike skeletal muscle, cardiac muscle has a relatively *long* refractory period, and is thus incapable of summation. In fact, cardiac muscle is incapable of reacting to any stimulus before about the middle of phase 3, and will not respond to a normal cardiac stimulus before phase 4. The period of time between the beginning of the cardiac action potential and the approximate middle of phase 3 is the **effective refractory period.** The period of time between the effective refractory period and phase 4 is the **relative refractory period.** The refractory period of cardiac muscle is about 250 milliseconds—almost as long as the contraction of the heart.

In this activity we will use direct heart stimulation to better understand the refractory period of cardiac muscle.

1.   Click on the **Direct Heart Stimulation** electrode and drag it to the electrode holder (the post with the four green squares.) Release the electrode; it will click into place, touching the ventricle of the heart.

2.   Look at the heartbeat trace. Recall that the smaller rise is atrial contraction and the larger rise is ventricular contraction. Click **Single Stimulus** right at the beginning of ventricular contraction.

Did you see any change in the trace?

_____

Why/why not?

_____

3.   Click **Single Stimulus** again, this time near the peak of ventricular contraction.

Did you see any change in the trace?

_____

Why/why not?

_____

4.   Click **Single Stimulus** again, this time during the "fall" of ventricular contraction. If you do not see any change in the trace, try clicking **Single Stimulus** twice in quick succession until you do see a change.

Describe the change you see in the trace. How does it differ from the baseline trace?

_____

_____

_____

5.   Next, click on **Multiple Stimuli** (note that the button changes to **Stop Stimulus**). This will administer repeated stimuli (20 per second) to the heart. After a few seconds, click **Stop Stimulus** to stop the stimulations.

What effect do the repeated stimuli have on the heartbeat trace? Describe the trace.

_____

_____

_____

What part of the trace shows the refractory period?

_____

Does summation occur?

_____

Why is it important that summation not occur in heart muscle?

_____

_____

_____

_____

6.   Return the **Direct Heart Stimulation** electrode to its original location. ■

Activity 2:
Vagus Nerve Stimulation

In this activity we will be indirectly stimulating the heart by stimulating the vagus nerve. Recall that the vagus nerve carries parasympathetic signals to the heart.

1.   Click the **Vagus Nerve Stimulation** electrode and drag it to the electrode holder. Notice that when the electrode clicks into place, the vagus nerve will be draped over it. Any stimuli applied will go directly to the vagus nerve and indirectly to the heart.

2.   Click **Multiple Stimulus.** Note that the number of stimuli is preset to 50 stimuli/second.

3.   Observe the effects of stimulation on the heartbeat trace. Allow the trace to cross the length of the monitor five times before you click **Stop Stimulus.**

Describe the effect of the vagus nerve stimulation on the heartbeat.

_____

_____

_____

Soon after you applied the stimuli, did the heart rate increase or decrease?

_____

Why?

_____

_____

How do the sympathetic and parasympathetic nervous systems work together to regulate heart rate?

_____

_____

_____

Complete the following: As heart rate decreases, cardiac output _____. As heart rate increases, cardiac output _____.

Research has shown that in the absence of neural or hormonal influences, the SA node generates action potentials at a frequency of approximately 100 beats per minute. Yet, the average resting heart rate is 70 beats per minute. What does this tell you about the parasympathetic nervous system in relation to the sympathetic nervous system and hormones?

_____

_____

_____ ■

Modifiers of Heart Rate

Next we will observe the effects of various drugs, hormones, and ions on heart rate. Click **Experiment** at the top of the screen and then select **Modifiers of Heart Rate.** You will see the opening screen shown in Figure 6.4. Notice the seven dropper bottles resting on a shelf above the monitor. You will be administering drops of each bottle's contents onto the frog heart and observing the effects. You will also be keeping a record of your data by clicking the **Record Data** button on the data collection box at the bottom of the screen. You may print your recorded data at any time by clicking **Tools** at the top of the screen and then selecting **Print Data.**

Activity 3:
Effect of Epinephrine

Epinephrine is a hormone secreted by the adrenal glands. It travels through the bloodstream to the heart, on which it has a significant effect. Epinephrine is an important mediator of rapid fuel mobilization, or increasing metabolic rate.

1.   Click the top of the dropper bottle labeled **Epinephrine** and drag the dropper over to the top of the frog heart and release the mouse button. You will see drops of epinephrine released onto the heart.

2.   Observe the heart rate trace and watch the status window on the lower right corner of the heart rate monitor.

3.   Wait until the status window reads _Heart Rate Stable,_ then click **Record Data.**

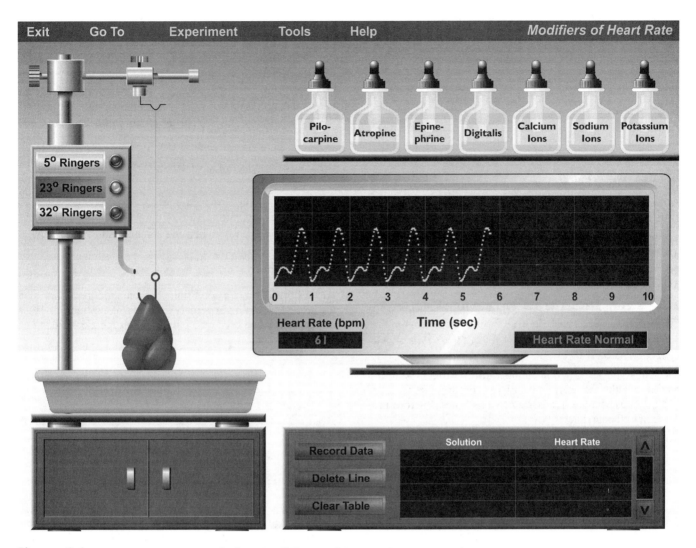

Figure 6.4 Opening screen of the Modifiers of Heart Rate experiment.

4. Click the **23°C** (room temperature) **Ringers** solution button to bathe the heart and flush out the epinephrine. Wait until the status window reads *Heart Rate Normal* before proceeding to the next activity.

What was the effect of epinephrine on heart rate?

_____

Does this effect mimic the effect of the sympathetic nervous system or that of the parasympathetic nervous system?

_____

What is the heart rate when the status window reads *Heart

Rate Stable*? _____ ■

A c t i v i t y   4 :

Effect of Pilocarpine

Pilocarpine is a *cholinergic* drug—that is, a drug that inhibits, mimics, or enhances the action of acetylcholine in the body.

1. Click the top of the dropper bottle labeled **Pilocarpine** and drag the dropper over to the top of the frog heart and release the mouse button. You will see drops of pilocarpine released onto the heart.

2. Observe the heart rate trace and watch the status window on the lower right corner of the heart rate monitor.

3. Wait until the status window reads *Heart Rate Stable,* then click **Record Data.**

4. Click the **23°C** (room temperature) **Ringers** solution button to bathe the heart and flush out the pilocarpine. Wait until the status window reads *Heart Rate Normal* before proceeding to the next activity.

What was the effect of pilocarpine on heart rate?

_____

What is the heart rate when the status window reads _Heart_

Rate Stable? _____ ▪

Activity 5:
Effect of Atropine

Atropine is another cholinergic drug, although its effects are quite different from that of pilocarpine.

1. Click the top of the dropper bottle labeled **Atropine** and drag the dropper over to the top of the frog heart and release the mouse button. You will see drops of atropine released onto the heart.

2. Observe the heart rate trace and watch the status window on the lower right corner of the heart rate monitor.

3. Wait until the status window reads _Heart Rate Stable,_ then click **Record Data.**

4. Click the **23°C** (room temperature) **Ringers** solution button to bathe the heart and flush out the atropine. Wait until the status window reads _Heart Rate Normal_ before proceeding to the next activity.

What was the effect of atropine on heart rate?

_____

What is the heart rate when the status window reads _Heart_

Rate Stable? _____

Circle the correct boldfaced term: Atropine _inhibits / mimics / enhances_ the action of acetylcholine on heart rate.

If you were to administer drops of pilocarpine to the heart and then administer atropine immediately afterward, what effect would you expect to see on the heart rate?

_____

_____ ▪

Activity 6:
Effect of Digitalis

Digitalis is a drug that interferes with the normal conduction pathway in the heart by blocking the conduction of the atrial impulses to the ventricles.

1. Click the top of the dropper bottle labeled **Digitalis** and drag the dropper over to the top of the frog heart and release the mouse button. You will see drops of digitalis released onto the heart.

2. Observe the heart rate trace and watch the status window on the lower right corner of the heart rate monitor.

3. Wait until the status window reads _Heart Rate Stable,_ then click **Record Data.**

4. Click the **23°C** (room temperature) **Ringers** solution button to bathe the heart and flush out the digitalis. Wait until the status window reads _Heart Rate Normal_ before proceeding to the next activity.

What was the effect of digitalis on heart rate?

_____

What is the heart rate when the status window reads _Heart_

Rate Stable? _____

Atrial fibrillation is a condition characterized by the atria of the heart undergoing extremely fast rates of contraction. Why might digitalis be used on a patient with such a condition?

_____

_____

_____ ▪

Activity 7:
Effect of Temperature

The frog is a _poikilothermic_ animal; its internal body temperature changes depending on the temperature of its external environment. In contrast, humans are _homeothermic._ Our bodies remain at the same temperature (around 37°C) unless we are sick.

1. The frog heart is currently being bathed in room temperature Ringers solution. Click the **5° Ringers** solution button to observe the effects of lowering the temperature.

2. Observe the heart rate trace and watch the status window on the lower right corner of the heart rate monitor.

3. Wait until the status window reads _Heart Rate Stable,_ then click **Record Data.**

4. Click the **23° Ringers** solution button to bathe the heart and return it to room temperature. Wait until the status window reads _Heart Rate Normal_ before proceeding.

5. Now click on the **32° Ringers** solution button to observe the effects of raising the temperature.

6. Observe the heart rate trace and watch the status window on the lower right corner of the heart rate monitor.

7. Wait until the status window reads _Heart Rate Stable,_ then click **Record Data.**

8. Click the **23° Ringers** solution button to bathe the heart and return it to room temperature. Wait until the status window reads _Heart Rate Normal_ before continuing to the next activity.

What was the effect of temperature on the frog's heart rate?

_____

_____

_____ ■

Activity 8:
Effects of Ions

A large amount of potassium is present inside the cardiac muscle cell. Sodium and calcium are present in larger quantities outside the cell. The resting cell membrane favors the conduction of potassium over both sodium and calcium. This means that the resting membrane potential of cardiac cells is determined mainly by the ratio of extracellular and intracellular concentrations of potassium. You may wish to review the section on the five phases of cardiac action potential at the beginning of this lab on page 55 before proceeding with this activity, in which you will be examining the effects of calcium, sodium, and potassium ions on the frog heart rate.

1. Click the top of the dropper bottle labeled **Calcium Ions** and drag the dropper over to the top of the frog heart and release the mouse button. You will see drops of calcium ions released onto the heart.

2. Observe the heart rate trace and watch the status window on the lower right corner of the heart rate monitor.

3. Wait until the status window reads _Heart Rate Stable,_ then click **Record Data.**

4. Click the **23°C** (room temperature) **Ringers** solution button to bathe the heart and flush out the calcium. Wait until the status window reads _Heart Rate Normal_ before proceeding.

5. Repeat steps 1–4 with **Sodium Ions** and then **Potassium Ions.**

What was the effect of calcium ions on heart rate?

_____

Why?

_____

_____

Where in a heart cell is calcium normally found?

_____

What was the effect of sodium ions on heart rate?

_____

Why?

_____

_____

Where in a heart cell is sodium normally found?

_____

What was the effect of potassium ions on heart rate?

_____

Why?

_____

_____

Where in a heart cell is potassium normally found?

_____ ■

Remember, you may print your recorded data at any time by clicking **Tools** at the top of the screen and then clicking **Print Data.**

Histology Review Supplement

Turn to p. 140 for a review of cardiovascular tissue.

7

Respiratory System Mechanics

Objectives

1. To explain how the respiratory and circulatory systems work together to enable gas exchange among the lungs, blood, and body tissues

2. To define respiration, ventilation, alveoli, diaphragm, inspiration, expiration, and partial pressure

3. To explain the differences between tidal volume, inspiratory reserve volume, expiratory reserve volume, vital capacity, residual volume, total lung capacity, forced vital capacity, forced expiratory volume, and minute respiratory volume

4. To list various factors that affect respiration

5. To explain how surfactant works in the lungs to promote respiration

6. To explain what happens in pneumothorax

7. To explain how hyperventilation, rebreathing, and breathholding affect respiratory volumes

The physiological functions of respiration and circulation are essential to life. If problems develop in other physiological systems, we can still survive for some time without addressing them. But if a persistent problem develops within the respiratory or circulatory systems, death can ensue within minutes.

The primary role of the respiratory system is to distribute oxygen to, and remove carbon dioxide from, the cells of the body. The respiratory system works hand in hand with the circulatory system to achieve this. The term **respiration** includes *breathing*—the movement of air in and out of the lungs, also known as **ventilation**—as well as the transport (via blood) of oxygen and carbon dioxide between the lungs and body tissues. The heart pumps deoxygenated blood to pulmonary capillaries, where gas exchange occurs between blood and **alveoli** (air sacs in the lungs), oxygenating the blood. The heart then pumps the oxygenated blood to body tissues, where oxygen is used for cell metabolism. At the same time, carbon dioxide (a waste product of metabolism) from body tissues diffuses into the blood. The deoxygenated blood then returns to the heart, completing the circuit.

Ventilation is the result of muscle contraction. The **diaphragm**—a dome-shaped muscle that divides the thoracic and abdominal cavities—contracts, making the thoracic cavity larger. This reduces the pressure within the thoracic

cavity, allowing atmospheric gas to enter the lungs (a process called **inspiration**). When the diaphragm relaxes, the pressure within the thoracic cavity increases, forcing air out of the lungs (a process called **expiration**). Inspiration is considered an "active" process because muscle contraction requires the use of ATP, whereas expiration is usually considered a "passive" process. When a person is running, however, the *external intercostal muscles* contract and make the thoracic cavity even larger than with diaphragm contraction alone, and expiration is the result of the *internal intercostal muscles* contracting. In this case, both inspiration and expiration are considered "active" processes, since muscle contraction is needed for both. Intercostal muscle contraction works in conjunction with diaphragm muscle contraction.

Respiratory Volumes

Ventilation is measured as the frequency of breathing multiplied by the volume of each breath, called the **tidal volume.** Ventilation is needed to maintain oxygen in arterial blood and carbon dioxide in venous blood at their normal levels—that is, at their normal **partial pressures.** [The term *partial pressure* refers to the proportion of pressure that a single gas exerts within a mixture. For example, in the atmosphere at sea level, the pressure is 760 mm Hg. Oxygen makes up about 20% of the total atmosphere and therefore has a partial pressure (P_{O_2}) of 760 mm Hg $\times$ 20%, close to 160 mm Hg.]

Oxygen diffuses down its partial pressure gradient to flow from the alveoli of the lungs into the blood, where the oxygen attaches to hemoglobin (meanwhile, carbon dioxide diffuses from the blood to the alveoli). The oxygenated blood is then transported to body tissues, where oxygen again diffuses down its partial pressure gradient to leave the blood and enter the tissues. Carbon dioxide (produced by the metabolic reactions of the tissues) diffuses down its partial pressure gradient to flow from the tissues into the blood for transport back to the lungs. Once in the lungs, the carbon dioxide follows its partial pressure gradient to leave the blood and enter the air in the alveoli for export from the body.

Normal tidal volume in humans is about 500 milliliters. If one were to breathe in a volume of air equal to the tidal volume and then continue to breathe in as much air as possible, that amount of air (above and beyond the tidal volume) would equal about 3100 milliliters. This amount of air is called the **inspiratory reserve volume.** If one were to breathe *out* as much air as possible beyond the normal tidal volume, that amount of air (above and beyond the tidal volume) would equal about 1200 milliliters. This amount of air is called the **expiratory reserve volume.** Tidal volume, inspiratory reserve volume, and expiratory reserve volume together consti-

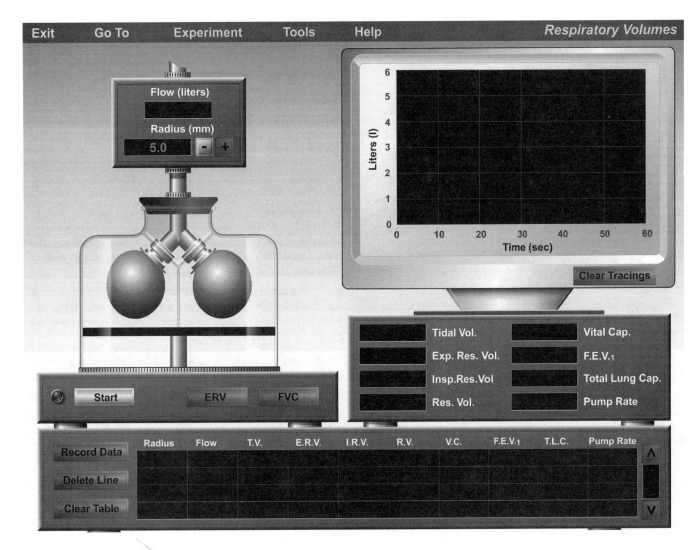

Figure 7.1 Opening screen of the Respiratory Volumes experiment.

tute the **vital capacity,** about 4800 milliliters. It is important to note that the histological structure of the respiratory tree (where air is found in the lungs) will not allow *all* air to be breathed out of the lungs. The air remaining in the lungs after a complete exhalation is called the **residual volume,** normally about 1200 milliliters. Therefore, the **total lung capacity** (the vital capacity volume plus the residual volume) is approximately 6000 milliliters.

All of these volumes can be easily measured using a *spirometer.* Basically, a spirometer is composed of an inverted bell in a water tank. A breathing tube is connected to the bell's interior. On the exterior of the inverted bell is attached a pen device that records respiratory volumes on paper. When one exhales into the breathing tube, the bell goes up and down with exhalation. Everything is calibrated so that respiratory volumes can be read directly from the paper record. The paper moves at a pre-set speed past the recording pen so that volumes per unit time can be easily calculated. In addition to measuring the respiratory volumes introduced so far, the spirometer can also be used to perform pulmonary function tests. One such test is the **forced vital capacity**

(FVC), or the amount of air that can be expelled completely and as rapidly as possible after taking in the deepest possible breath. Another test is the **forced expiratory volume (FEV$_1$),** which is the percentage of vital capacity that is exhaled during a 1-sec period of the FVC test. This value is generally 75% to 85% of the vital capacity.

In the following experiments you will be simulating spirometry and measuring each of these respiratory volumes using a pair of mechanical lungs. Follow the instructions in the "Getting Started" section at the front of this lab manual to start up PhysioEx. From the Main Menu, select **Respiratory System Mechanics.** You will see the opening screen for the "Respiratory Volumes" experiment (Figure 7.1). At the left is a large vessel (simulating the thoracic cavity) containing an *air flow tube.* This tube looks like an upside-down "Y." At the ends of the "Y" are two spherical containers, simulating the lungs, into which air will flow. On top of the vessel are controls for adjusting the radius of the tube feeding the "lungs." This tube simulates the trachea and other air passageways into the lungs. Beneath the "lungs" is a black platform simulating the diaphragm. The "diaphragm" will move down, simulating contraction and increasing the volume of the "thoracic cavity" to bring air into the "lungs"; it will then move up, simulating relaxation and decreasing the volume of the "thoracic cavity" to expel air out. At the bottom of the vessel are three buttons: a **Start** button, an **ERV** (expiratory reserve volume) button, and an **FVC** (forced vital capacity) button. Clicking **Start** will start the simulated lungs breathing at normal tidal volume; clicking **ERV** will cause the lungs to exhale as much air as possible beyond the tidal volume; and clicking **FVC** will cause the lungs to expel the most air possible after taking the deepest possible inhalation.

At the top right is an oscilloscope monitor, which will graphically display the respiratory volumes. Note that the *Y*-axis displays liters instead of milliliters. The *X*-axis displays elapsed time, with the length of the full monitor displaying 60 seconds. Below the monitor is a series of data displays. A data recording box runs along the bottom length of the screen. Clicking **Record Data** after an experimental run will record your data for that run on the screen.

Activity 1:
Trial Run

Let's conduct a trial run to get familiarized with the equipment.

1. Click the **Start** button (notice that it immediately turns into a **Stop** button.) Watch the trace on the oscilloscope monitor, which currently displays normal tidal volume. Watch the simulated diaphragm rise and fall, and notice the "lungs" growing larger during inhalation and smaller during exhalation. The **Flow** display on top of the vessel tells you the amount of air (in liters) being moved in and out of the lungs with each breath.

2. When the trace reaches the right side of the oscilloscope monitor, click the **Stop** button and then click **Record Data.** Your data will appear in the data recording box along the bottom of the screen. This line of data tells you a wealth of information about respiratory mechanics. Reading the data from left to right, the first data field should be that of the *Radius* of the air flow tube (5.00 mm). The next data field, *Flow,* displays the total flow volume for this experimental run. *T.V.* stands for "Tidal Volume"; *E.R.V.* for "Expiratory Reserve Volume"; *I.R.V.* for "Inspiratory Reserve Volume"; *R.V.* for "Residual Volume"; *V.C.* for "Vital Capacity"; FEV_1 for "Forced Expiratory Volume"; *T.L.C.* for "Total Lung Capacity"; and finally, *Pump Rate* for the number of breaths per minute.

3. You may print your data at any time by clicking **Tools** at the top of the screen and then **Print Data.** You may also print the trace on the oscilloscope monitor by clicking **Tools** and then **Print Graph.**

4. Highlight the line of data you just recorded by clicking it and then click **Delete Line.**

5. Click **Clear Tracings** at the bottom right of the oscilloscope monitor. You are now ready to begin the first experiment.

■

Activity 2:
Measuring Normal Respiratory Volumes

1. Make sure that the radius of the air flow tube is at 5.00 mm. To adjust the radius, click the (+) or (−) buttons next to the radius display.

2. Click the **Start** button. Watch the oscilloscope monitor. When the trace reaches the 10-second mark on the monitor, click the **ERV** button to obtain the expiratory reserve volume.

3. When the trace reaches the 30-second mark on the monitor, click the **FVC** to obtain the forced vital capacity.

4. Once the trace reaches the end of the screen, click the **Stop** button, then click **Record Data.**

5. Remember, you may print your trace or your recorded data by clicking **Tools** at the top of the screen and selecting either **Print Graph** or **Print Data.**

From your recorded data, you can calculate the **minute respiratory volume:** the amount of air that passes in and out of the lungs in 1 minute. The formula for calculating minute respiratory volume is:

**Minute respiratory volume =
tidal volume × bpm (breaths per minute)**

Click **Tools** and then **Calculator.** Calculate and enter the minute respiratory volume: _____

Judging from the trace you generated, inspiration took place over how many seconds? _____

Expiration took place over how many seconds?

_____

Does the duration of inspiration or expiration vary during ERV or FVC? _____

6. Click **Clear Tracings** before proceeding to the next activity. Do *not* delete your recorded data—you will need it for the next activity. ■

Activity 3:
Effect of Restricted Air Flow on Respiratory Volumes

1. Adjust the radius of the air flow tube to 4.00 mm by clicking the (−) button next to the radius display. Repeat steps 2–5 from the previous activity, making sure to click **Record Data.**

How does this set of data compare to the data you recorded for Activity 2?

_____

_____

Is the respiratory system functioning better or worse than it did in the previous activity? Explain why.

_____

_____

2. Click **Clear Tracings.**
3. Reduce the radius of the air flow tube by another 0.50 mm to 3.50 mm.
4. Repeat steps 2–6 from Activity 2.
5. Reduce the radius of the air flow tube by another 0.50 mm to 3.00 mm.
6. Repeat steps 2–6 from Activity 2.

What was the effect of reducing the radius of the air flow tube on respiratory volumes?

_____

_____

What does the air flow tube simulate in the human body?

_____

_____

What could be some possible causes of reduction in air flow to the lungs?

_____

_____

7. Click **Tools → Print Data** to print your data.

Express your FEV_1 data as a percentage of vital capacity by filling out the following chart. (That is, take the FEV_1 value and divide it into the vital capacity value for each line of data.)

FEV_1 as % of Vital Capacity

Radius	FEV_1	Vital Capacity	FEV_1 (%)
5.00			
4.00			
3.50			
3.00			

Factors Affecting Respiration

Many factors affect respiration. *Compliance,* or the ability of the chest wall or lung to distend, is one. If the chest wall or lungs cannot distend, respiratory ability will be compromised. **Surfactant,** a lipid material secreted into the alveolar fluid, is another. Surfactant acts to decrease the surface tension of water in the fluid that lines the walls of the alveoli. Without surfactant, the surface tension of water would cause alveoli to collapse after each breath. A third factor affecting respiration is any injury to the thoracic wall that results in the wall being punctured. Such a puncture would effectively raise the intrathoracic pressure to that of atmospheric pressure, preventing diaphragm contraction from decreasing intrathoracic pressure and, consequently, preventing air from being drawn into the lungs. (Recall that airflow is achieved by the generation of a pressure difference between atmospheric pressure on the outside of the thoracic cavity and intrathoracic pressure on the inside.)

We will be investigating the effect of surfactant in the next activity. Click **Experiment** at the top of the screen and then select **Factors Affecting Respiration.** The opening screen will look like Figure 7.2. Notice the changes to the equipment above the air flow tube. Clicking the **Surfactant** button will add a pre-set amount of surfactant to the "lungs." Clicking **Flush** will clear the lungs of surfactant. Also notice that valves have been added to the sides of each simulated lung. Opening the valves will allow atmospheric pressure into the vessel (the "thoracic cavity"). Finally, notice the changes to the display windows below the oscilloscope screen. **Flow Left** and **Pressure Left** refer to the flow of air and pressure in the left "lung"; **Flow Right** and **Pressure Right** refer to the flow of air and pressure in the right "lung." **Total Flow** is the sum of Flow Left and Flow Right.

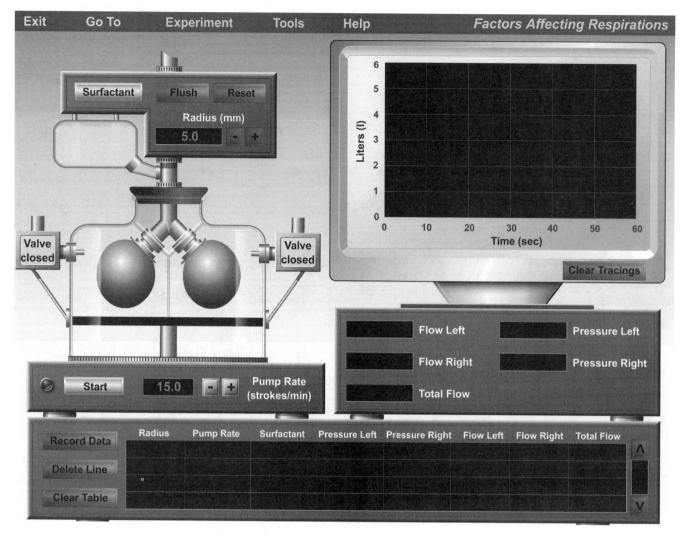

Figure 7.2 Opening screen of the Factors Affecting Respiration experiment.

Activity 4:
Effect of Surfactant on Respiratory Volumes

1. The data recording box at the bottom of the screen should be clear of data. If not, click **Clear Table.**

2. The radius of the air flow tube should be set at 5.0 mm, and the Pump Rate should be set at 15 strokes/minute.

3. Click **Start** and allow the trace to sweep across the full length of the oscilloscope monitor. Then click **Record Data.** This will serve as the baseline, or control, for your experimental runs. You may wish to click **Tools** and then **Print Graph** for a printout of your trace.

4. Click **Surfactant** twice to add surfactant to the system. Repeat step 3.

When surfactant is added, what happens to the tidal volume?

_____

As a result of the tidal volume change, what happens to the flow into each lung and total air flow?

_____

Why does this happen?

_____

_____

Remember, you may click **Tools** and then either **Print Data** or **Print Graphs** to print your results. ■

Activity 5:
Effect of Thoracic Cavity Puncture

Recall that if the wall of the thoracic cavity is punctured, the intrathoracic pressure will equalize with atmospheric pressure so that the lung cannot be inflated. This condition is known as **pneumothorax,** which we will investigate in this next activity.

1. Do *not* delete your data from the previous activity.

2. If there are any tracings on the oscilloscope monitor, click **Clear Tracings.**

3. Click **Flush** to remove the surfactant from the previous activity.

4. Be sure that the air flow radius is set at 5.0 mm, and that Pump Rate is set at 15 strokes/minute.

5. Click on **Start** and allow the trace to sweep the length of the oscilloscope monitor. Notice the pressure displays, and how they alternate between positive and negative values.

6. Click **Record Data.** Again, this is your baseline data.

7. Now click the valve for the left lung, which currently reads "Valve closed."

8. Click **Start** and allow the trace to sweep the length of the oscilloscope monitor.

9. Click **Record Data.**

What happened to the left lung when you clicked on the valve button? Why?

_____

_____

_____

What has happened to the "Total Flow" rate?

_____

What is the pressure in the left lung? _____

Has the pressure in the right lung been affected? _____

If there was nothing separating the left lung from the right lung, what would have happened when you opened the valve for the left lung? Why?

_____

_____

_____

Now click the valve for the left lung again, closing it. What happens? Why?

_____

_____

Click **Reset** (next to the **Flush** button at the top of the air flow tube.) What happened?

_____

_____

Describe the relationship required between intrathoracic pressure and atmospheric pressure in order to draw air into the lungs.

_____

_____

Design your own experiment for testing the effect of opening the valve of the right lung. Was there any difference from the effect of opening the valve of the left lung?

_____

Remember, you may click **Tools** and then either **Print Data** or **Print Graphs** to print your results. ■

Variations in Breathing

Normally, alveolar ventilation keeps pace with the needs of body tissues. The adequacy of alveolar ventilation is measured in terms of the partial pressure of carbon dioxide (P_{CO_2}). Carbon dioxide is the major component for regulating breathing rate. Ventilation (the frequency of breathing multiplied by the tidal volume) maintains the normal partial pressures of oxygen and carbon dioxide both in the lungs and blood. *Perfusion,* the pulmonary blood flow, is matched to ventilation. The breathing patterns of an individual are tightly regulated by the breathing centers of the brain so that the respiratory and circulatory systems can work together effectively.

In the next activity you will examine the effects of rapid breathing, rebreathing, and breathholding on the levels of carbon dioxide in the blood. Rapid breathing increases breathing rate and alveolar ventilation becomes excessive for tissue needs. It results in a decrease in the ratio of carbon dioxide production to alveolar ventilation. Basically, alveolar ventilation becomes too great for the amount of carbon dioxide being produced. In rebreathing, air is taken in that was just expired, so the P_{CO_2} (the partial pressure of carbon dioxide) in the alveolus (and subsequently in the blood) is elevated. In breathholding, there is no ventilation and no gas exchange between the alveolus and the blood.

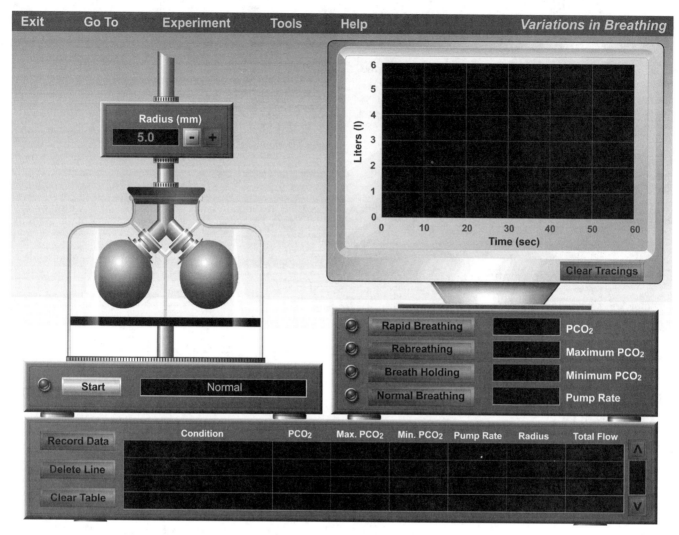

Figure 7.3 Opening screen of the Variations in Breathing experiment.

Click **Experiment** at the top of the screen and select **Variations in Breathing.** You will see the next screen, shown in Figure 7.3. This screen is very similar to the ones you have been working on. Notice the buttons for **Rapid Breathing, Rebreathing, Breath Holding,** and **Normal Breathing**—clicking each of these buttons will induce the given pattern of breathing. Also note the displays for P_{CO_2}, Maximum P_{CO_2}, Minimum P_{CO_2}, and **Pump Rate.** ■

Activity 6:
Rapid Breathing

1. The oscilloscope monitor and the data recording box should both be empty and clear. If not, click **Clear Tracings** or **Clear Table.**

2. The air flow tube radius should be set to 5.0. If not, click the (+) or (−) buttons next to the radius display to adjust it.

3. Click **Start** and conduct a baseline run. Remember to click **Record Data** at the end of the run. Leave the baseline trace on the oscilloscope monitor.

4. Click **Start** again, but this time click the **Rapid Breathing** button when the trace reaches the 10-second mark on the oscilloscope monitor. Observe the P_{CO_2} levels in the display windows.

5. Allow the trace to finish, then click **Record Data.**

What happens to the P_{CO_2} level during rapid breathing?

_____

Why?

_____

_____

Remember, you may click **Tools** and then either **Print Data** or **Print Graphs** to print your results.

Click **Clear Tracings** before continuing to the next activity. ■

Activity 7:
Rebreathing

Repeat Activity 6, except this time click the **Rebreathing** button instead of the **Rapid Breathing** button.

What happens to the P_{CO_2} level during rebreathing?

_____

_____

Why?

_____

_____

Did the total flow change?

_____

Why?

_____

_____

How does the rebreathing trace compare to your baseline trace? (Look carefully—differences may be subtle.)

_____

Why?

_____

_____

Click **Clear Tracings** to clear the oscilloscope monitor. ◼

Activity 8:
Breath Holding

1.   Click on **Start** and conduct a baseline run. Remember to click **Record Data** at the end of the run. Leave the baseline trace on the oscilloscope monitor.

2.   Click **Start** again, but this time click the **Breath Holding** button when the trace reaches the 10-second mark on the oscilloscope monitor. Observe the P_{CO_2} levels in the display windows.

3.   At the 20-second mark, click **Normal Breathing** and let the trace finish.

4.   Click **Record Data.**

What happens to the P_{CO_2} level during rebreathing?

_____

Why?

_____

_____

What change was seen when you returned to "Normal Breathing"?

_____

_____ ◼

Remember, you may print your data or graphs by clicking **Tools** at the top of the screen and then selecting either **Print Data** or **Print Graph.**

Histology Review Supplement

Turn to p. 141 for a review of respiratory tissue.

Chemical and Physical Processes of Digestion

Objectives

1. To define **digestive tract, accessory glands, digestion, hydrolases, salivary amylase, carbohydrates, proteins, lipids, bile salts, pepsin,** and **lipase**

2. To understand the main functions and processes of the digestive system

3. To understand the specificity of enzyme action

4. To explain the impact of temperature and pH levels on enzyme activity

5. To identify the three main categories of food molecules

6. To explain how enzyme activity can be assessed with enzyme assays

7. To identify the main enzymes, substrates, and products of carbohydrate, protein, and fat digestion

The digestive system, also called the *gastrointestinal system,* consists of the **digestive tract** (also *gastrointestinal tract* or *GI tract*) and **accessory glands** that secrete enzymes and fluids needed for digestion. The digestive tract includes the mouth, pharynx, esophagus, stomach, small intestine, colon, rectum, and anus. The major functions of the digestive system are to ingest food, to break food down to its simplest components, to extract nutrients from these components for absorption into the body, and to expel wastes.

Most of the food we consume cannot be absorbed into our bloodstream without first being broken down into smaller particles. **Digestion** is the process of breaking down food molecules into smaller molecules with the aid of enzymes in the digestive tract. Digestive enzymes are **hydrolases:** They *catalyze* (accelerate) the addition of water to food molecules in order to break them down into smaller subunits. For example, when two amino acids join together to form a protein, an —OH$^-$ group is removed from the carboxyl end of one amino acid, and an —H$^+$ is removed from the amino group of the second amino acid to form a dipeptide bond between the two amino acids plus water. To break down such a protein, a digestive enzyme catalyzes the addition of water (—OH$^-$ plus —H$^+$) to the dipeptide bond, cleaving the bond to restore the carboxyl group of the first amino acid and the amino group of the second amino acid, and effectively breaks the protein down into two amino acid subunits. Once a food

molecule is broken down into its simplest components, the components are absorbed through the epithelial cells that line the intestinal tract and then enter the bloodstream.

In addition to being hydrolases, digestive enzymes are *substrate specific*—they work on some substances but not others. For example, **salivary amylase** is an enzyme in the saliva that breaks down starch (found in foods like corn, potatoes, bread, and pasta) and glycogen (animal starch), but not cellulose (found in the cell walls of plants), even though cellulose is made up of glucose, just like starch and glycogen.

Two factors that play key roles in the efficacy of digestive enzymes are temperature and pH level. An increase in temperature may cause a reaction to speed up, as it causes molecules to move faster and thus increases contact with an enzyme; however, too high a temperature will disrupt molecular bonding that stabilizes enzyme configuration, causing the enzyme to *denature* (i.e., undergo a structural change that renders it functionless). In addition, each enzyme has an optimal pH at which it is most active. Within range of this optimal pH, the enzyme will work as expected; beyond the optimal pH, the enzyme may have no effect.

Most food molecules fall into one of the following categories: carbohydrates, proteins, or lipids. **Carbohydrates** are the principal source of calories for most people and include glucose, sugars, and starch. Larger carbohydrates are broken down into *monosaccharides* (simple sugars, such as glucose) before being absorbed into the blood. **Proteins** are very important for growth, especially among young people. Proteins are broken down into *amino acids* before being absorbed into the body to build new proteins. **Lipids,** most of which are *triglycerides* (the major constituents of fats and oils), are not water soluble and thus pose special problems for digestion. Lipase, the enzyme that acts on lipids, is hydrolytic (like all digestive enzymes) and can only work on the surfaces of lipid droplets because the lipids are water insoluble. To increase the rate of digestion by lipase, lipids are first *emulsified* (broken down into smaller droplets) with the aid of **bile salts,** a cholesterol derivative. Emulsification results in smaller droplets with larger surface areas, making it easier for lipase to bind to substrates and digest lipids. Bile salts also form **micelles,** which aid in the absorption of products of lipid digestion: *fatty acids* and *monoglycerides.*

In the following experiments you will be examining the effects of different digestive enzymes on carbohydates, proteins, and lipids. To begin, follow the instructions for starting up PhysioEx 3.0 in the "Getting Started" section at the beginning of this lab manual. From the Main Menu, select **Chemical and Physical Processes of Digestion.** The opening screen will appear (see Figure 8.1). This screen will be used for the first two activities, in which you will be testing the effects of salivary amylase on starch and cellulose.

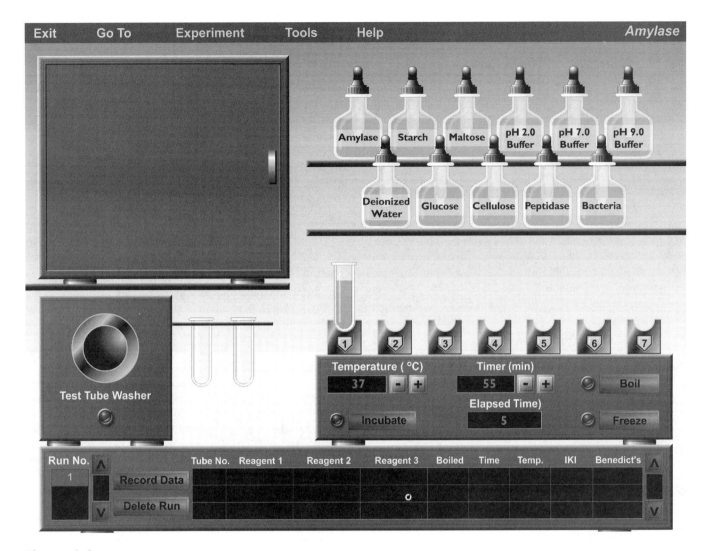

Figure 8.1 Opening screen of the Amylase experiment.

Amylase

Notice the 11 dropper bottles in the top right quadrant of the screen. You will be preparing test tubes containing various combinations of the dropper bottle contents. Below the dropper bottles is an incubation unit that will enable you to boil, freeze, and incubate the test tubes. The closed cabinet in the upper left quadrant of the screen is an *assay cabinet,* containing chemicals that you will add to your experimental test tubes in order to interpret your test results. Below the assay cabinet is a *test tube washer* where you will deposit used test tubes. Next to this is a *test tube dispenser* from which you will click test tubes and drag them to the holders in the incubation unit, where you will prepare them for experimentation. Along the bottom of the screen is the data recording box, where you will be recording your experimental data.

A c t i v i t y   1 :
Salivary Amylase and Starch

Starch digestion begins in the mouth with the action of salivary amylase, which is secreted by salivary glands. Salivary amylase breaks starch down into *maltose,* a disaccharide formed from two molecules of glucose. Thus the presence of maltose in an experimental sample would indicate that starch digestion has occurred.

Your goals in this experiment are to test the effects of amylase on starch, determine the optimal pH level at which amylase works, and observe the effects of temperature on enzyme activity.

1. Click the test tube hanging nearest to the incubator, drag it to slot number 1 on top of the incubator, and release the mouse button. The test tube will click into place. Repeat this action, filling slot number 2, slot number 3, and so on, until all seven slots on top of the incubator contain test tubes.

2. Fill your seven test tubes with three substances each, as follows:

> Test tube #1: starch, deionized water, pH 7.0 buffer
> Test tube #2: amylase, deionized water, pH 7.0 buffer
> Test tube #3: starch, amylase, pH 7.0 buffer
> Test tube #4: starch, amylase, pH 7.0 buffer
> Test tube #5: maltose, deionized water, pH 7.0 buffer
> Test tube #6: starch, amylase, pH 2.0 buffer
> Test tube #7: starch, amylase, pH 9.0 buffer

To do this, click the dropper cap of the desired substance, drag it to the top of the desired test tube, and release the mouse button.

3. Click the number "3" under tube #3. The tube will lower into the incubation unit. Then click Boil. The tube will boil and then resurface. Note that the only difference between tube #3 and tube #4 is that tube #3 is being boiled.

4. Be sure the incubation temperature is set for 37°C and that the timer is set for 60 minutes. If not, click the (+) or (−) buttons accordingly.

5. Click **Incubate.** All seven test tubes will be lowered into the incubation unit and gently agitated while incubated. At the end of the incubation period, the tubes will resurface and the door of the assay cabinet will open.

You will notice that there are seven empty test tubes and two dropper bottles in the assay cabinet. The two dropper bottles contain **IKI** and **Benedict's** reagents. IKI tests for the presence of starch, while Benedict's tests for the presence of maltose (which, you will recall, is the product of starch digestion). You will be adding these reagents to your seven experimental test tubes to determine whether or not digestion has taken place.

6. Click test tube #1 in slot 1 of the incubator. You will see your cursor change into a miniature test tube. Drag this miniature test tube to the rim of the first tube in the assay cabinet, then release the mouse button. The contents of the miniature test tube will empty into the assay tube.

7. Repeat step 6 for the remaining test tubes. Be sure to do this in sequence; that is, do test tube #2, then test tube #3, and so on.

8. Once solutions from each of the tubes on the incubator have been transferred to the tubes in the assay cabinet, click the top of the IKI dropper bottle, drag the dropper to the opening of the first tube in the assay cabinet, and release the mouse button. Drops of IKI will be dispensed. Repeat this for all the test tubes in the assay cabinet and note any color changes. A *blue-black* or *gray* color indicates a positive test for starch; a yellow color indicates a negative test. Record your IKI results in the following chart.

9. Next, click the dropper top for Benedict's, drag it to the opening of test tube #1, (in slot 1 of the incubator), and release the mouse button. Drops of Benedict's will be added to the test tube. Repeat this for the remaining test tubes on top of the incubator.

10. After Benedict's reagent has been added to each test tube on top of the incubator, click **Boil.** All the test tubes will descend into the incubator unit, boil, and resurface. Examine the tubes for color changes. A green, orange, or reddish color indicates the presence of maltose, considered a positive Benedict's result. A blue color indicates that no maltose is present, and is considered a negative Benedict's result. Record your data in Chart 1.

Chart 1   Results of Activity 1

Tube No.	1	2	3	4	5	6	7
Additives	starch deionized water pH 7.0 buffer	amylase deionized water pH 7.0 buffer	starch amylase pH 7.0 buffer	starch amylase pH 7.0 buffer	maltose deionized water pH 7.0 buffer	starch amylase pH 2.0 buffer	starch amylase pH 9.0 buffer
Incubation condition	37°C	37°C	boiled, then incubated at 37°C	37°C	37°C	37°C	37°C
IKI test							
Benedict's test							

11. Click **Record Data** to record your data onscreen. You may also click **Tools** and select **Print Data** for a hard-copy print out of your results.

12. Click and drag each test tube to the opening of the test tube washer and release the mouse button. The tubes will disappear.

What was the purpose of tubes #1 and #2?

_____

_____

What can you conclude from tubes #3 and #4?

_____

_____

What do tubes #4, #6, and #7 tell you about amylase activity and pH levels?

_____

_____

_____

_____

What is the optimal pH for amylase activity? _____

Does amylase work at pH levels other than the optimal pH?

_____

What is the end-product of starch digestion?

_____

In which tubes did you detect the presence of maltose at the end of the experiment?

_____

Why wasn't maltose present in the other tubes?

_____

_____

Salivary amylase would be greatly deactivated in the stomach. Suggest a reason why, based on what you have learned in this activity.

_____

_____

Activity 2:
Salivary Amylase and Cellulose

If there are any test tubes remaining in the incubator, click and drag them to the test tube washer before beginning this activity.

In the previous activity we learned that salivary amylase can digest starch. In this activity we will test to see whether or not salivary amylase digests cellulose, a substance found in the cell walls of plants. We will also investigate whether bacteria (such as that found in our large intestine) will digest cellulose, and whether peptidase (a pancreatic enzyme that breaks down peptides) will digest starch.

1. Click the test tube hanging nearest to the incubator, drag it to slot number 1 on top of the incubator, and release the mouse button. The test tube will click into place. Repeat this action, filling slot number 2, slot number 3, and so on, until all seven slots on top of the incubator contain test tubes.

2. Fill your seven test tubes with three substances each, as follows:

Test tube #1: amylase, starch, pH 7.0 buffer
Test tube #2: amylase, starch, pH 7.0 buffer
Test tube #3: amylase, glucose, pH 7.0 buffer
Test tube #4: amylase, cellulose, pH 7.0 buffer
Test tube #5: amylase, cellulose, deionized water
Test tube #6: peptidase, starch, pH 7.0 buffer
Test tube #7: bacteria, cellulose, pH 7.0 buffer

To do this, click the dropper cap of the desired substance, drag it to the top of the desired test tube, and release the mouse button.

3. Click the number "1" under tube #1. The tube will lower into the incubation unit. Then click **Freeze.** The tube will be frozen and then resurface.

4. Be sure the incubation temperature is set for 37°C and that the timer is set for 60 minutes. If not, click the (+) or (−) buttons accordingly.

5. Click **Incubate.** All seven test tubes will be lowered into the incubation unit and gently agitated while incubated. At the end of the incubation period, the tubes will resurface and the door of the assay cabinet will open.

Again, you will notice that there are seven empty test tubes and the IKI and Benedict's dropper bottles in the assay cabinet. Recall that IKI tests for the presence of starch, while Benedict's tests for the presence of maltose. You will be adding these reagents to your seven experimental test tubes to determine whether or not digestion has taken place.

6. Click test tube #1 in slot 1 of the incubator. You will see your cursor change into a miniature test tube. Drag this miniature test tube to the rim of the first tube in the assay cabinet, then release the mouse button. The contents of the miniature test tube will empty into the assay tube.

7. Repeat step 6 for the remaining test tubes. Be sure to do this in sequence; that is, do test tube #2, then test tube #3, and so on.

8. Once solutions from each of the tubes on the incubator have been transferred to the tubes in the assay cabinet, click the top of the IKI dropper bottle, drag the dropper to the

opening of the first tube in the assay cabinet, and release the mouse button. Drops of IKI will be dispensed. Repeat this for all the test tubes in the assay cabinet and note any color changes. A *blue-black* or *gray* color indicates a positive test for starch; a yellow color indicates a negative test. Record your IKI results in the following chart.

9. Next, click dropper top for Benedict's, drag it to the opening of test tube #1 (in slot 1 of the incubator), and release the mouse button. Drops of Benedict's will be added to the test tube. Repeat this for the remaining test tubes on top of the incubator.

10. After Benedict's reagent has been added to each test tube on top of the incubator, click **Boil.** All the test tubes will descend into the incubator unit, boil, and resurface. Examine the tubes for color changes. A green, orange, or reddish color indicates the presence of maltose, considered a positive Benedict's result. A blue color indicates that no maltose is present, and is considered a negative Benedict's result. Record your data in Chart 2.

11. Click **Record Data** to record your data onscreen. You may also click **Tools** and select **Print Data** for a hard copy print out of your results.

12. Click and drag each test tube to the opening of the test tube washer and release the mouse button. The tubes will disappear.

Which tubes showed a positive test for the IKI reagent?

_____

Which tubes showed a positive test for Benedict's reagent?

_____

What was the effect of freezing tube #1?

_____

_____

How does the effect of freezing differ from the effect of boiling?

_____

_____

What is the smallest subunit into which starch can be broken down?

_____

_____

What was the effect of amylase on glucose in tube #3? Suggest an explanation for this effect.

_____

_____

Chart 2 Results of Activity 2

Tube No.	1	2	3	4	5	6	7
Additives	amylase starch pH 7.0 buffer	amylase starch pH 7.0 buffer	amylase glucose pH 7.0 buffer	amylase cellulose pH 7.0 buffer	amylase cellulose deionized water	peptidase starch pH 7.0 buffer	bacteria cellulose pH 7.0 buffer
Incubation condition	frozen, then incubated at 37°C	37°C	37°C	37°C	37°C	37°C	37°C
IKI test							
Benedict's test							

What was the effect of amylase on cellulose in tube #4?

_____

_____

Popcorn and celery are nearly pure plant starch or cellulose. What can you conclude about the digestion of cellulose, judging from the results of test tubes #4, 5, and 7?

_____

_____

What was the effect of the enzyme peptidase, used in tube #6? Explain your answer, based on what you know about peptidase and substrate-specificity.

_____

_____ ■

Pepsin

Proteins are composed of subunits known as *amino acids.* When subjected to enzyme activity, proteins are broken down to their amino acid components. **Pepsin** is an example of an enzyme that breaks down protein. Pepsin is secreted by stomach glands as an inactive proenzyme, *pepsinogen,* which is converted to pepsin by the cleavage of acid-labile linkages in the acidic (low pH) environment of the stomach. The extent to which protein in the stomach is hydrolyzed or digested is significant but variable. It is estimated that 15% of dietary protein is reduced to amino acids by pepsin. Most protein digestion occurs in the duodenum of the small intestine.

Activity 3:
Pepsin Digestion of Protein

In this next activity you will be investigating the effects of pepsin on BAPNA, a synthetic protein. Click **Experiment** at the top of the screen and then select **Pepsin.** You will see the screen shown in Figure 8.2. The screen is slightly different

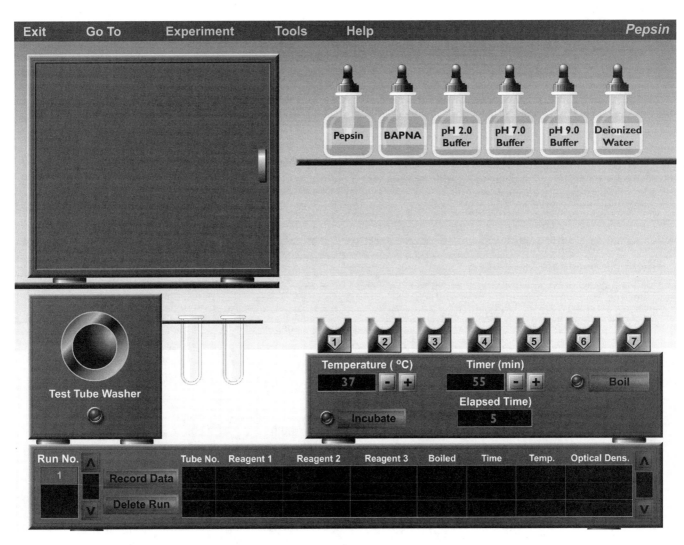

Figure 8.2 Opening screen of the Pepsin experiment.

from the one you worked with for the "Amylase" activities. Notice that the dropper bottles now include bottles for pepsin and for BAPNA. BAPNA is colorless and transparent in solution but will become yellow if digested by an enzyme like pepsin—you will not need to add additional reagents to determine whether or not enzyme activity has occurred.

1. Click the test tube hanging nearest to the incubator, drag it to slot number 1 on top of the incubator, and release the mouse button. The test tube will click into place. Repeat this action, filling slot number 2, slot number 3, and so on, until *six* of the seven slots on top of the incubator contain test tubes. (Unlike the previous activities, you will be working with only six test tubes here.)

2. Fill your six test tubes with three substances each, as follows:

Test tube #1: pepsin, BAPNA, pH 2.0 buffer
Test tube #2: pepsin, BAPNA, pH 2.0 buffer
Test tube #3: pepsin, deionized water, pH 2.0 buffer
Test tube #4: deionized water, BAPNA, pH 2.0
Test tube #5: pepsin, BAPNA, pH 7.0 buffer
Test tube #6: pepsin, BAPNA, pH 9.0 buffer

To do this, click the dropper cap of the desired substance, drag it to the top of the desired test tube, and release the mouse button.

3. Click the number "1" under tube #1. The tube will lower into the incubation unit. Then click **Boil.** The tube will boil and then resurface.

4. Be sure the incubation temperature is set for 37°C and that the timer is set for 60 minutes. If not, click the (+) or (−) buttons accordingly.

5. Click **Incubate.** All six test tubes will be lowered into the incubation unit and gently agitated while incubated. At the end of the incubation period, the tubes will resurface and the door of the assay cabinet will open.

When the assay cabinet opens, you will see a *spectrophotometer,* an instrument that measures the amount of light absorbed (called *optical density*) by a solution. You will be using the spectrophotometer to measure how much yellow dye was released when BAPNA was "digested." To do this, you will individually drag each test tube into the spectrophotometer and click **Analyze.** The spectrophotometer will shine a light through the solution to measure its optical density. The greater the optical density, the more BAPNA digestion by pepsin occurred.

6. Click tube #1, drag it to the spectrophotometer, and release the mouse button.

7. Click **Analyze.**

8. Note the optical density reading for this test tube. Record it in the chart below.

9. Remove the test tube and return it to its slot on top of the incubator.

10. Repeat steps 6–9 for the remaining test tubes.

11. After all of the tubes have been read, click **Record Data** to record your data on screen.

12. Drag each test tube to the test tube washer to dispose of them.

13. Click **Tools → Print Data** to print your data.

Which pH value allowed for maximal hydrolysis of BAPNA?

_____

What effect did the boiling have on enzyme activity?

_____

Chart 3   Results of Activity 3

Tube No.	1	2	3	4	5	6
Additives	pepsin BAPNA pH 2.0 buffer	pepsin BAPNA pH 2.0 buffer	pepsin deionized water pH 2.0 buffer	deionized water BAPNA pH 2.0 buffer	pepsin BAPNA pH 7.0 buffer	pepsin BAPNA pH 9.0 buffer
Incubation condition	boiled, then incubated at 37°C	37°C	37°C	37°C	37°C	37°C
Optical density						

Which test tubes led you to this conclusion?

_____

Would freezing have the same effect? Why or why not?

_____

Which test tubes were your "controls"?

_____

What were they "controlling" for?

_____

What was pepsin's affect on BAPNA?

_____

Using the simulation, design an experiment that would allow you to test how the amount of BAPNA digestion varies with time. What is your conclusion?

_____

_____

Using the simulation, design an experiment that would allow you to test whether or not temperature has any affect on BAPNA digestion. What is your conclusion?

_____

_____ ■

Lipase

In a normal diet, the primary lipids are triglycerides, the major constituents of fats and oils. Lipids are insoluble in water, and thus they must first be emulsified (broken down into smaller droplets, thus increasing their surface areas) before a digestive enzyme like **lipase** can work effectively on them. In the small intestine, lipids are emulsified by bile, a yellow-green fluid produced by the liver. The resulting bile-covered droplets have relatively large surface areas, allowing water-soluble lipase enzymes easier access to substrates. Lipase then hydrolzes the lipid droplets to fatty acids and monoglyc-erides, the end-products of lipid digestion. Hydrolysis of lipids also forms micelles, small molecular aggregates that increase absorption of the products of lipid digestion. Lipases found in pancreatic juice are responsible for the digestion of most of the lipids present in normal diets.

Activity 4:
Lipase, Bile, and Lipid Digestion

Click **Experiment** at the top of the screen and then select **Lipase.** You will see the screen shown in Figure 8.3. Notice that the dropper bottles now include lipase, vegetable oil, and bile salts. Notice also that the dropper bottles containing pH buffers are now color-coded—this will help you analyze your results later in the experiment. You will be testing the effects of lipase and bile on the digestion of a lipid: vegetable oil.

1.   Click the test tube hanging nearest to the incubator, drag it to slot number 1 on top of the incubator, and release the mouse button. The test tube will click into place. Repeat this action, filling slot number 2, slot number 3, and so on, until all seven slots on top of the incubator contain test tubes.

2.   Fill your seven test tubes with four substances each, as follows:

Test tube #1: lipase, vegetable oil, bile salts, pH 7.0 buffer
Test tube #2: lipase, vegetable oil, bile salts, pH 7.0 buffer
Test tube #3: lipase, vegetable oil, deionized water, pH 7.0 buffer
Test tube #4: lipase, deionized water, bile salts, pH 9.0 buffer
Test tube #5: deionized water, vegetable oil, bile salts, pH 7.0 buffer
Test tube #6: lipase, vegetable oil, bile salts, pH 2.0 buffer
Test tube #7: lipase, vegetable oil, bile salts, pH 9.0 buffer

To do this, click the dropper cap of the desired substance, drag it to the top of the desired test tube, and release the mouse button.

3.   Click the number "1" under tube #1. The tube will lower into the incubation unit. Then click **Boil.** The tube will boil and then resurface.

4.   Be sure the incubation temperature is set for 37°C and that the timer is set for 60 minutes. If not, click the (+) or (−) buttons accordingly.

5.   Click **Incubate.** All seven test tubes will be lowered into the incubation unit and gently agitated while incubated. At the end of the incubation period, the tubes will resurface and the door of the assay cabinet will open.

You will see a pH meter in the assay cabinet. Digestion of vegetable oil by lipase will release fatty acids, which decrease pH levels. Thus you will be using the pH meter to help you detect the presence of fatty acids (and, consequently, evidence of digestion) by comparing the pH levels of your samples now to their original pH levels. You will be clicking and dragging the test tubes individually into the pH meter. Once the tubes lock into place, you will click **Measure pH.** An electrode will descend into the tube's contents and take a pH reading.

6.   Click and drag test tube #1 into the pH meter.

7.   Click **Measure pH.**

8.   Record the pH value in Chart 4 on page 78, then return the test tube to its slot on the incubator.

9.   Repeat steps 6–8 for the remaining test tubes.

10.   When all of the tubes have been analyzed, click **Record Data.**

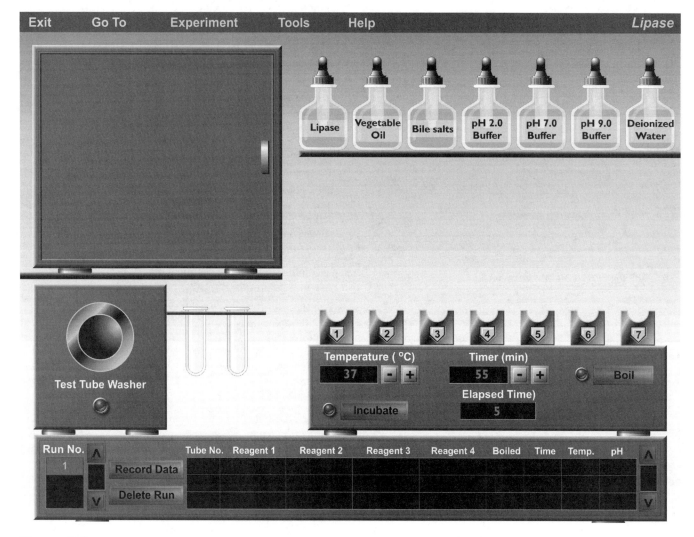

Figure 8.3 Opening screen of the Lipase experiment.

Chart 4   Results of Activity 4

Tube No.	1	2	3	4	5	6	7
Additives	lipase vegetable oil bile salts pH 7.0 buffer	lipase vegetable oil bile salts pH 7.0 buffer	lipase vegetable oil deionized water pH 7.0 buffer	lipase deionized water bile salts pH 9.0 buffer	deionized water vegetable oil bile salts pH 7.0 buffer	lipase vegetable oil bile salts pH 2.0 buffer	lipase vegetable oil bile salts pH 9.0 buffer
Incubation condition	boiled, then incubated at 37°C	37°C	37°C	37°C	37°C	37°C	37°C
pH							

11. Click **Tools** at the top of the screen and then **Print Data** if you wish to print out a hard copy of your data.

12. Dispose of the test tubes by dragging them individually to the test tube washer.

What did tube #1 show?

_____

Why was tube #1 found to have a pH of 7?

_____

What is the main difference between tubes #2 and #3?

_____

_____

What is the action of the substance in tube #2 that was not included in tube #3?

_____

_____

What was the optimal pH level for lipase digestion?

_____

How do you know?

_____

What effect did the other buffers have on the digestive process?

_____

_____

What are the products of lipase digestion in tube #2?

_____

In what tubes did you detect the presence of fatty acids?

_____

_____

Physical Processes of Digestion

Keep in mind that the chemical processes of enzyme activity are only part of the digestive process. Physical processes are also involved. Chewing, for example, mixes food with salivary mucus and amylase, reduces it to small, manageable particles and transforms it into a swallowable mass called a *bolus.* The tongue separates the bolus from the mass of food in the mouth, pressing it against the hard palate. The bolus then enters the pharynx, stimulating tactile receptors that initiate the swallowing reflex and propel the bolus through the esophagus. A wave of contraction called *peristalsis* then moves the bolus into the stomach, where the bolus is converted to *chyme.* Stomach muscles mix the chyme with gastric juices to fragment the food into still smaller particles; they also regulate the entry of chyme into the small intestine. Peristalsis continues in the small intestine, periodically interspersed by *segmentation,* in which the chyme is shuttled back and forth by the contraction and relaxation of the intestinal segments. Segmentation thoroughly mixes chyme with digestive enzymes, bile, and bicarbonate ions secreted by the pancreatic duct, and increases the efficiency of nutrient absorption into the blood.

Histology Review Supplement

Turn to p. 142 for a review of digestive tissue.

Objectives

1. To define **nephron, renal corpuscle, renal tubule, afferent arteriole, glomerular filtration, efferent arteriole, aldosterone, ADH,** and **reabsorption**

2. To describe the components and functions of a nephron

3. To understand how arterial diameter affects nephron function

4. To understand how blood pressure affects nephron function

5. To explain the process of reabsorption

6. To explain the role of carriers in glucose reabsorption

7. To understand the actions of ADH and aldosterone on solute reabsorption and water uptake

The kidneys are excretory and regulatory organs. By excreting water and solutes, the kidneys are responsible for ridding the body of waste products and excess water. The kidneys regulate 1) plasma *osmolarity,* or the concentration of a solution expressed as osmoles of solute per liter of solvent; 2) plasma volume; 3) acid-base balance; 4) electrolyte balance; 5) excretion of metabolic wastes and foreign materials; and 6) the production and secretion of hormones that regulate osmolarity and electrolyte balance. All these activities are extremely important to maintaining homeostasis in the body.

The kidneys are located between the posterior abdominal wall and the abdominal peritoneum. Although many textbooks depict the kidneys directly across from each other, the right kidney is actually slightly lower than the left. Each human kidney contains approximately 1.2 million **nephrons,** the functional units of the kidney. Each nephron is composed of a **renal corpuscle** and a **renal tubule.** The renal corpuscle consists of a tuft of capillaries, called the *glomerulus,* which is enclosed by a fluid-filled capsule called *Bowman's capsule.* An **afferent arteriole** supplies blood to the glomerulus. As blood flows through the glomerular capillaries, protein-free plasma filters into the Bowman's capsule, a process called **glomerular filtration.** An **efferent arteriole** then drains the glomerulus of the remaining blood. The filtrate flows from Bowman's capsule to the start of the renal tubule, called the *proximal convoluted tubule,* then on to the

proximal straight tubule, followed by the *loop of Henle,* a U-shaped hairpin loop. The filtrate then flows into the *distal convoluted tubule* before reaching the *connecting tubule* and the *collecting duct,* where urine collects. The distal tubule and collecting duct are composed of two cell types: *principal cells* and *intercalated cells.* Principal cells reabsorb Na^+ and water and secrete K^+. Intercalated cells secrete either H^+ or HCO_3^- and are, therefore, very important in the regulation of the acid-base balance.

Glomerular Filtration

Let's take a closer look at what happens during glomerular filtration. Blood enters the glomerulus from the afferent arteriole. *Starling forces* (hydrostatic and osmotic pressure gradients) drive protein-free plasma from the blood across the walls of the glomerular capillaries and into the Bowman's capsule. The *glomerular filtration rate* is an index of kidney function. In humans, the filtration rate ranges from 80 to 140 ml/min, so that in 24 hours as much as 180 liters of plasma is filtered by glomeruli. The filtrate formed is devoid of cellular debris and is essentially protein free. The concentration of salts and organic molecules are similar to that of blood. Normal urine output is 1–1.5 liters/24 hours. The difference is reabsorbed in the body. Normally, only about 20% of the blood that enters the nephron is filtered, due to the osmotic pressure of the blood *(oncotic pressure)* and the hydrostatic pressure from the fluids in Bowman's capsule. The glomerular filtration rate can be altered by changing afferent arteriole resistance, efferent arteriole pressure, or the size of the filtration surface, or by a process called renal autoregulation.

Once the filtrate is formed, the nephron must reabsorb materials that the body needs and excrete unneeded materials from the body. While as much as 180 liters are filtered each day, less than 1% of the filtered water, sodium chloride, and other solutes are excreted in the urine. More than 67% of this reabsorption takes place in the proximal convoluted tubule. The distal convoluted tubule and collecting duct reabsorb approximately 7% of the filtered NaCl, secrete a variable amount of K^+ and H^+, and reabsorb a variable amount of water. It is in this distal part of the nephron that hormones act to reabsorb water and electrolytes. **Aldosterone** regulates NaCl reabsorption (and thus NaCl excretion as well). **ADH** (antidiuretic hormone) causes the permeability of the distal tubule and collecting duct to increase, promoting the uptake of water from the filtrate. ADH is considered the body's most important hormone for regulating water balance.

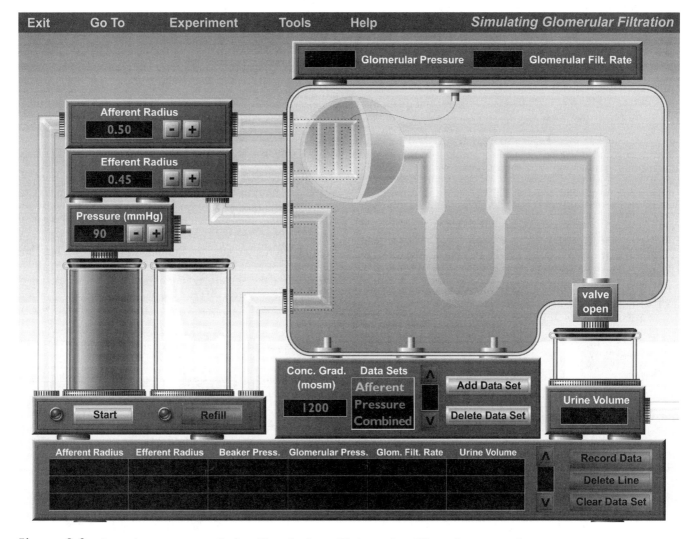

Figure 9.1 **Opening screen of the Simulating Glomerular Filtration experiment.**

In the first three activities you will be concentrating on how arterial diameter and pressure affect glomerular filtration rate and urine volume. Follow the instructions for starting PhysioEx in the "Getting Started" section at the front of this manual. From the Main Menu, select **Renal System Physiology.** You will see the screen shown in Figure 9.1.

Click **Help** at the top of the screen and then select **Balloons On.** Now move your mouse around the simulated nephron in the yellow section of the screen. Labels will appear for the various parts of the nephron as you roll over them. Note in particular the glomerulus and the glomerulus capsule. Also note the "afferent tube" and "efferent tube" to the left of the glomerulus—these represent the afferent and efferent arterioles delivering and draining blood from the glomerulus. You may adjust the radius of either of these tubes by clicking the (+) and (−) buttons next to the respective tubes. You may also adjust the blood pressure of the source beaker by clicking the (+) and (−) buttons next to the "Pressure (mmHg)" display.

Once you have identified all the equipment on screen, click **Help** again and select **Balloons Off** (you cannot proceed with the experiment unless the labels are turned off). At the bottom left of the screen are two beakers. The left beaker, which we call the "source beaker," represents the blood supply being delivered to the nephron. When the **Start** button is clicked, blood will flow from the source beaker to the afferent arteriole and then to the group of small tubes representing the glomerulus. As blood flows through the glomerulus, you will see *ultrafiltration* occur. (Ultrafiltration means filtration from the plasma of everything except proteins and cells.) Blood will then be drained from the glomerulus to the "drain beaker" next to the source beaker. At the end of the nephron tube, you will see the formation of urine in a small beaker at the lower right of the screen. To watch a test run of this process in action, click the **Start** button. At the end of the run, click **Refill** underneath the drain beaker before you begin the activities that follow.

Activity 1:
Effect of Arteriole Diameter on Glomerular Filtration

In this activity you will investigate how the diameters of the afferent and efferent arterioles leading to and from the glomerulus can affect the glomerular filtration rate.

1.   The afferent radius display should be set at 0.50 mm, and the efferent radius at 0.45 mm. If not, use the (**+**) or (**−**) buttons next to the afferent and efferent radius displays to adjust accordingly.

2.   Be sure the left beaker is full. If not, click the **Refill** button.

3.   The pressure gauge above the left beaker should read 90 mm Hg. If not, click the (**+**) or (**−**) buttons next to the pressure display to adjust accordingly.

4.   Click the **Start** button. As the blood flows through the nephron, watch the displays for glomerular pressure and glomerular filtration rate at the top right of the screen, as well as the display for urine volume at the bottom right of the screen.

5.   After the drain beaker has stopped filling with blood, click **Record Data.** This will be your baseline data for this activity.

6.   Click the **Refill** button.

7.   Increase the *afferent* radius by 0.05 mm and repeat steps 3–6, making sure to click **Record Data** at the end of each run. Keep all the other variables at their original settings. Continue repeating the activity until you have reached the maximum afferent radius of 0.60 mm.

Compare this data with your baseline data. How did increasing the afferent arteriole radius affect glomerular filtration rate?

_____

8.   Reduce the afferent arteriole radius to 0.30 mm, and click **Start.**

Under these conditions, does the fluid flow through the nephron?

_____

What is the glomerular filtration rate? _____

How does it compare to your baseline data, and why?

_____

_____

9.   Using the simulation, design and carry out an experiment for testing the effects of increasing or decreasing the *efferent* radius.

How did increasing the efferent radius affect glomerular filtration rate?

_____

_____

How did decreasing the efferent radius affect glomerular filtration rate?

_____

_____

Physiologically, what could be the cause of a change in afferent or efferent arteriole radius?

_____

_____

Activity 2:
Effect of Pressure on Glomerular Filtration

Next you will investigate the effect of blood pressure on glomerular filtration rate.

1. Under the **Data Sets** display, highlight **Pressure.** This will allow you to save data in a new data set window. You can always retrieve your data from the previous activity by highlighting the **Afferent** data set.

2. Make sure that the source beaker is filled with blood, and that the drain beaker is empty. If not, click **Refill.**

3. Adjust the pressure gauge (on top of the source beaker) to 70 mm Hg. Set the afferent radius at 0.50 mm and the efferent radius at 0.45 mm.

4. Click the **Start** button. Watch the Glomerular Pressure and Glomerular Filtration Rate displays at the top right of the screen.

5. When the run has finished, click the **Record Data** button. This is your baseline data.

6. Increase the pressure by 5 mm Hg and repeat the experiment. Continue increasing the pressure by 5 mm and repeating the experiment until you have reached the maximum pressure of 100 mm Hg. Be sure to click **Record Data** and **Refill** after each experimental run.

As pressure increased, what happened to the pressure in the glomerulus?

_____

_____

What happened to the glomerular filtration rate?

_____

Compare the urine volume in your baseline data with the urine volume as you increased the pressure.

How did the urine volume change?

_____

How could increased urine volume be viewed as being beneficial to the body?

_____

_____

Activity 3:
Combined Effects

In the first activity you looked at arteriole diameter and its role in glomerular filtration. Next you examined the effect of pressure on glomerular filtration. In the human body, both of these effects are occurring simultaneously. In this activity you will investigate the combined effects of arteriole diameter and pressure changes on glomerular filtration.

1. Under the **Data Sets** window, highlight **Combined.** This will allow you to save data in a new data set window. You can always retrieve your data from previous activities by highlighting the **Afferent** data set or the **Pressure** data set.

2. Set the pressure at 90 mm Hg, the afferent arteriole at 0.50 mm, and the efferent arteriole at 0.45 mm.

3. Click the **Start** button and allow the run to complete. Then click **Record Data.** This is your baseline data.

4. Click **Refill.**

5. Lower the pressure to 80 mm Hg. Leave afferent arteriole at 0.50 and efferent arteriole at 0.45 mm.

6. Click the **Start** button and allow the run to complete. Then click **Record Data.**

7. Click **Refill.**

What happened to the glomerular filtration rate and urine volume after you reduced the pressure?

_____

_____

How could you adjust the afferent or efferent radius to compensate for the effect of the reduced pressure on glomerular filtration rate and urine volume? Use the simulation to determine your answer.

_____

_____

8. Next, click the square valve button (currently reading "valve open") above the collecting duct. The valve should now read "valve closed."

9. Click **Start.** At the end of the run, click **Record Data.**

What changes are seen in nephron function when the valve is closed?

_____

_____

Why were these changes seen?

_____

_____

_____

_____

Is the kidney functional when the glomerular filtration rate is zero? Explain your answer.

_____

_____

What is the major "ingredient" that needs to be removed from the blood?

_____

Studies on aging have demonstrated that some nephrons may fail as we get older. Will this be a problem regarding urine formation?

_____

_____

If blood pressure went down—for example, as the result of blood loss—what changes would the kidney need to make to maintain its normal filtration rate?

_____

_____

10. Click **Tools** → **Print Data** to print your data.

Simulating Urine Formation

Reabsorption is the movement of filtered solutes and water from the lumen of the renal tubules back into the plasma. Without reabsorption, we would excrete the solutes and water that our bodies need. In the next activity you will examine the process of passive reabsorption that occurs while filtrate travels through a nephron and urine is formed.

Click **Experiment** at the top of the screen and select **Simulating Urine Formation.** You will see the screen shown in Figure 9.2. The light yellow space surrounding the dark yellow nephron represents *interstitial space* between the nephron and *peritubular capillaries* that branch out from the

efferent arteriole. The movement of solutes and water from the renal tubules to the interstitial space is dependent on the *concentration gradient*—that is, the difference between the concentration of solutes in the tubules and the concentration of solutes in the interstitial space. Notice the display for **Conc. Grad. (mosm)** near the bottom of the screen. By clicking the (+) and (−) buttons next to this display and then clicking **Dispense,** you can adjust the solute concentration of the interstitial space. When you click **Start,** filtrate will begin flowing through the nephron, and solute and water will move from the tubules to the interstitial space to the peritubular capillaries, completing the process of reabsorption. Note that the capillaries are not shown on screen.

Note also the two dropper bottles at the right side of the screen, which contain the hormones aldosterone and ADH (antidiuretic hormone). For the first activity you will be dealing with ADH only, which increases the water permeability of the distal convoluted tube and the collecting duct of the nephron. Near the bottom left of the screen you will notice a **Probe.** When the probe turns red, you can click and drag it over various parts of the nephron and over the beaker to measure the solute concentration present. Finally, notice the equipment at the very top of the screen for adding glucose carriers. We will explain this equipment in Activity 5, when you will be studying the reabsorption of glucose.

Activity 4:
Effect of Solute Gradient on Urine Concentration

1. Make sure **Gradient** is highlighted within the **Data Sets** window.

2. Click and drag the dropper top from the bottle of ADH and release it on top of the gray cap directly above the right side of the nephron. The cap will open, and ADH will be dropped onto the collecting duct. The cap will then close.

3. The **Conc. Grad. (mosm)** window should read 300. If not, adjust the (+) or (−) buttons accordingly.

4. Click **Dispense** to apply the 300 mosm concentration to the interstitial fluid. It is important to note that this is also the typical value of solute concentration in the blood.

5. Click **Start** and allow the blood to flow through the system. Watch the **Probe** at the bottom left of the screen. When it turns red, click and drag it over to the urine collecting beaker to measure the urine solute concentration. The value will appear in the **Concentration** window next to the probe's original location.

6. At the end of the run, click **Record Data.**

7. Increase the concentration gradient by 300 mosm (i.e., set it at 600 mosm) and repeat the experiment. Remember to add ADH to the collecting duct before clicking **Start.** Continue increasing the gradient by 300 mosm and repeating the experiment until you reach 1200 mosm, the normal value for interstitial solute concentration in the kidney. Be sure to click **Record Data** after each run.

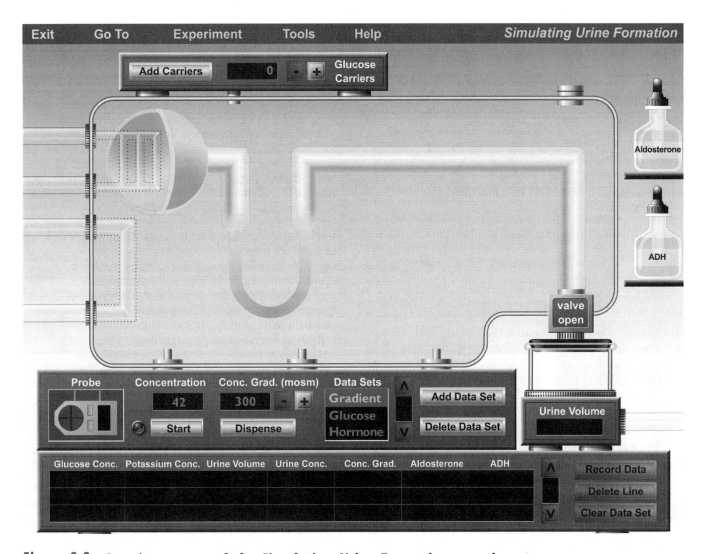

Figure 9.2 Opening screen of the Simulating Urine Formation experiment.

How did the urine solute concentration change as the concentration gradient of the interstitial fluid increased?

_____

_____

What happened to the urine volume as the concentration gradient increased? Why?

_____

_____

_____

By increasing the concentration gradient, what are you doing to the urine that is formed?

_____

_____

Predict what would happen to urine volume if you did not add ADH to the collecting duct.

_____

_____

Activity 5:
Reabsorption of Glucose

Glucose is not very large, and as such it is easily filtered out of the plasma into Bowman's capsule as part of the filtrate. To ensure that glucose is reabsorbed into the body so that it can form the starting material of metabolism, glucose carriers are present in the nephron. There is a finite number of carriers per cell, so if too much glucose is taken in, not all of it will be reabsorbed into the body. Glucose is absorbed by *secondary active transport,* the "push" of which comes from the gradient created by the transport of sodium. The carriers that transport these molecules across the membrane are proteins embedded in the cell membrane. In this activity you will examine the effect of glucose carriers on glucose reabsorption.

1.   Highlight **Glucose** within the **Data Sets** window.

2.   Set the **Conc. Grad. (mosm)** to 1200 and click **Dispense.** Recall that 1200 is the normal value for solute concentration in the kidney.

3.   Click **Start.**

4.   Click **Record Data** at the end of the run. This run will serve as a "control" run, with no glucose carriers present. Note that no ADH has been added, either—your focus will be on the absence or presence of glucose carriers.

5.   At the top of the screen, click the (**+**) button until the **Glucose Carriers** window reads 100. Then click **Add Carriers.**

6.   Click **Start.**

7.   At the end of the run, click **Record Data.**

8.   Continue to increase the number of glucose carriers by 100 at a time and repeat the experiment until you have reached the maximum number of carriers, 500. Be sure to dispense the concentration gradient before beginning each run, and to click **Record Data** after each run.

What happens to the glucose concentration as you add glucose carriers to the system?

_____

_____

At what point does the glucose concentration in the urine become zero?

_____

A person with type I diabetes cannot make insulin, and a person with type II diabetes does not respond to insulin that is made. In either case, the diabetic person is unable to absorb glucose into the body. What would you expect to find in the urine of a diabetic person? Why?

_____

_____

Activity 6:
Effect of Hormones on Reabsorption

By now you should understand how filtration occurs and how it is controlled. You should understand passive solute movement and the role of glucose carriers on glucose reabsorption. Next you will examine the actions of hormones on the nephron, most of which occur in the collecting duct.

Antidiuretic hormone (ADH) is influenced by the *osmolality* (the concentration of a solution expressed in osmoles of solute particles per kilogram of solvent) of body fluids as well as the volume and pressure of the cardiovascular system. A 1% change in body osmolality will cause this hormone to be secreted. The primary action of this hormone is to increase the permeability of the distal tubule and the collecting duct to water so that more water is taken up into the body. ADH binds to receptors in the principal cells to cause this reaction by opening *aquaporins,* or water channels in the apical membrane. Without water uptake, the body would quickly dehydrate. The kidney tightly regulates the amount of water excreted under normal conditions to maintain water balance in the body. Water intake must precisely match water loss from the body. If water intake is down, or if there has been a fluid loss from the body, the kidneys work to conserve water by making the urine very *hyperosmotic* (having a relatively high solute concentration) to the blood. If there has been a large intake of fluid, the urine is more *hypo-osmotic.* In the normal individual, urine osmolarity varies from 50 to 1200 milliosmoles/kg water. The osmolality of the body must be maintained within very narrow limits.

Aldosterone is an adrenal cortical hormone under the control of the body's *renin-angiotensin system.* A decrease in blood pressure is detected by cells in the afferent arteriole and triggers the release of renin. Renin acts as a proteolytic enzyme, causing angiotensinogen to be converted into angiotensin I. Endothelial cells have an enzyme, called the *converting enzyme,* which converts angiotensin I into angiotensin II. Angiotensin II works on the adrenal cortex to induce it to secrete aldosterone. Aldosterone acts on the collecting duct of the kidney to promote the uptake of sodium from filtrate *into* the body and the release of potassium *from* the body. Coupled with the addition of ADH, this electrolyte shift also causes more water to be reabsorbed into the blood, resulting in increased blood pressure.

1.   Highlight **Hormone** within the **Data Sets** window.

2.   Set the number of **Glucose Carriers** to zero and click **Dispense** to ensure that no carriers from the previous activity remain in action.

3.   Set the **Conc. Grad. (mosm)** to 1200 and click **Dispense.**

4.   Click **Start.** At the end of the run, click **Record Data.** This is your "control" run and baseline data.

5.   Click and drag the dropper top from the bottle of aldosterone and release it on top of the gray cap directly above the right side of the nephron. The cap will close, and aldosterone will be dropped onto the collecting duct.

6.   Click **Start.** At the end of the run, click **Record Data.**

7. Click and drag the dropper top from the bottle of ADH and release it on top of the gray cap directly above the right side of the nephron. The cap will close, and ADH will be dropped onto the collecting duct.

8. Click **Start.** At the end of the run, click **Record Data.**

9. For your fourth run, dispense both ADH and aldosterone onto the collecting duct. You should see a yellow outline appear around the collecting duct.

10. Click **Start.** At the end of the run, click **Record Data.**

11. Click **Tools → Print Data** to print your data.

Which hormone has the greater effect on urine volume? Why?

_____

_____

How does the addition of aldosterone affect the concentration of potassium in the urine?

_____

_____

How does the addition of ADH affect the concentration of potassium in the urine? How does this compare to the effect of adding aldosterone, with respect to potassium concentration in the urine?

_____

_____

How does the addition of both hormones affect 1) urine concentration, 2) urine volume, and 3) potassium concentration?

_____

_____

If ADH were not used, how would the urine concentration vary? Explain your answer.

_____

_____

Histology Review Supplement

Turn to p. 144 for a review of renal tissue.

Acid-Base Balance

Objectives

1. To define *pH* and identify the normal range of human blood pH levels.

2. To define *acid* and *base*, and explain what characterizes each of the following: *strong acid, weak acid, strong base, weak base.*

3. To explain how chemical and physiological buffering systems help regulate the body's pH levels.

4. To define the conditions of *acidosis* and *alkalosis.*

5. To explain the difference between *respiratory acidosis and alkalosis* and *metabolic acidosis and alkalosis.*

6. To understand the causes of respiratory acidosis and alkalosis.

7. To explain how the renal system compensates for respiratory acidosis and alkalosis.

8. To understand the causes of metabolic acidosis and alkalosis.

9. To explain how the respiratory system compensates for metabolic acidosis and alkalosis.

The term **pH** is used to denote the hydrogen ion concentration [H^+] in body fluids. pH values are the reciprocal of [H^+] and follow the formula

$$pH = \log(1/[H^+])$$

At a pH of 7.4, [H^+] is about 40 nanomolars (n*M*) per liter. Because the relationship is reciprocal, [H^+] is higher at *lower* pH values (indicating higher acid levels) and lower at *higher* pH values (indicating lower acid levels).

The pH of a body's fluids is also referred to as its **acid-base balance.** An **acid** is a substance that releases H^+ in solution (such as in body fluids). A **base,** often a hydroxyl ion (OH^-) or bicarbonate ion (HCO_3^-), is a substance that binds to H^+. A *strong acid* is one that completely dissociates in solution, releasing all of its hydrogen ions and thus lowering the solution's pH level. A *weak acid* dissociates incompletely and does not release all of its hydrogen ions in solution. A *strong base* has a strong tendency to bind to H^+, which has the effect of raising the pH value of the solution. A *weak base* binds less of the H^+, having a lesser effect on solution pH.

The body's pH levels are very tightly regulated. Blood and tissue fluids normally have pH values between 7.35 and 7.45. Under pathological conditions, blood pH values as low as 6.9 or as high as 7.8 have been recorded; however, values higher or lower than these cannot sustain human life. The narrow range of 7.35–7.45 is remarkable when one considers the vast number of biochemical reactions that take place in the body. The human body normally produces a large amount of H^+ as the result of metabolic processes, ingested acids, and the products of fat, sugar, and amino acid metabolism. The regulation of a relatively constant internal pH environment is one of the major physiological functions of the body's organ systems.

To maintain pH homeostasis, the body utilizes both *chemical* and *physiological* buffering sytems. Chemical buffers are composed of a mixture of weak acids and weak bases. They help regulate body pH levels by binding H^+ and removing it from solution as its concentration begins to rise, or releasing H^+ into solution as its concentration begins to fall. The body's three major chemical buffering systems are the *bicarbonate, phosphate,* and *protein buffer systems.* We will not focus on chemical buffering systems in this lab, but keep in mind that chemical buffers are the fastest form of compensation and can return pH to normal levels within a fraction of a second.

The body's two major physiological buffering systems are the renal and respiratory systems. The renal system is the slower of the two, taking hours to days to do its work. The respiratory system usually works within minutes, but cannot handle the amount of pH change that the renal system can. These physiological buffer systems help regulate body pH by controlling the output of acids, bases, or CO_2 from the body. For example, if there is too much acid in the body, the renal system may respond by excreting more H^+ from the body in urine. Similarly, if there is too much carbon dioxide in the blood, the respiratory system may respond by breathing faster to expel the excess carbon dioxide. Carbon dioxide levels have a direct effect on pH levels because the addition of carbon dioxide to the blood results in the generation of more H^+. The following reaction shows what happens in the respiratory system when carbon dioxide combines with water in the blood:

$$H_2O + CO_2 \rightleftarrows \underset{\substack{\text{carbonic} \\ \text{acid}}}{H_2CO_3} \rightleftarrows H^+ + \underset{\substack{\text{bicarbonate} \\ \text{ion}}}{HCO_3^-}$$

This is a reversible reaction and is useful for remembering the relationships between CO_2 and H^+. Note that as more CO_2 accumulates in the blood (which frequently is caused by reduced gas exchange in the lungs), the reaction moves to the right and more H^+ is produced, lowering the pH:

$$H_2O + \mathbf{CO_2} \rightarrow \underset{\substack{\text{carbonic} \\ \text{acid}}}{H_2CO_3} \rightarrow \mathbf{H^+} + \underset{\substack{\text{bicarbonate} \\ \text{ion}}}{HCO_3^-}$$

Conversely, as $[H^+]$ increases, more carbon dioxide will be present in the blood:

$$H_2O + \mathbf{CO_2} \leftarrow \underset{\substack{\text{carbonic} \\ \text{acid}}}{H_2CO_3} \leftarrow \mathbf{H^+} + \underset{\substack{\text{bicarbonate} \\ \text{ion}}}{HCO_3^-}$$

Disruptions of acid-base balance occur when the body's pH levels fall below or above the normal pH range of 7.35–7.45. When pH levels fall below 7.35, the body is said to be in a state of **acidosis.** When pH levels rise above 7.45, the body is said to be in a state of **alkalosis. Respiratory acidosis** and **respiratory alkalosis** are the result of the respiratory system accumulating too much or too little carbon dioxide in the blood. **Metabolic acidosis** and **metabolic alkalosis** refer to all other conditions of acidosis and alkalosis (i.e., those not caused by the respiratory system). The experiments in this lab will focus on these disruptions of acid-base balance, and on the physiological buffer systems (renal and respiratory) that compensate for such imbalances.

Respiratory Acidosis and Alkalosis

Respiratory acidosis is the result of impaired respiration, or *hypoventilation,* which leads to the accumulation of too much carbon dioxide in the blood. The causes of impaired respiration include airway obstruction, depression of the respiratory center in the brain stem, lung disease, and drug overdose. Recall that carbon dioxide acts as an acid by forming carbonic acid when it combines with water in the body's blood. The carbonic acid then forms hydrogen ions plus bicarbonate ions:

$$H_2O + \mathbf{CO_2} \rightarrow \underset{\substack{\text{carbonic} \\ \text{acid}}}{H_2CO_3} \rightarrow \mathbf{H^+} + \underset{\substack{\text{bicarbonate} \\ \text{ion}}}{HCO_3^-}$$

Because hypoventilation results in elevated carbon dioxide levels in the blood, the H^+ levels increase, and the pH value of the blood decreases.

Respiratory alkalosis is the condition of too little carbon dioxide in the blood. It is commonly the result of traveling to a high altitude (where the air contains less oxygen) or hyperventilation, which may be brought on by fever or anxiety. Hyperventilation removes more carbon dioxide from the blood, reducing the amount of H^+ in the blood and thus increasing the blood's pH level.

In this first set of activities, we focus on the causes of respiratory acidosis and alkalosis. Follow the instructions in the Getting Started section on pp. P-2 and P-3 in the PhysioEx Introduction to start PhysioEx. From the Main Menu, select **Acid-Base Balance.** You will see the opening screen for "Respiratory Acidosis/Alkalosis" (Figure 10.1). If you have already completed PhysioEx Exercise 7 on respiratory system mechanics, this screen should look familiar. At the left is a pair of simulated lungs, which look like balloons, connected by a tube that looks like an upside-down Y. Air flows in and out of this tube, which simulates the trachea and other air passageways into the lungs. Beneath the "lungs" is a black platform simulating the diaphragm. The long, U-shaped tube containing red fluid represents blood flowing through the lungs. At the top left of the U-shaped tube is a pH meter that will measure the pH level of the blood once the experiment is begun (experiments are begun by clicking the **Start** button at the left of the screen). To the right is an oscilloscope monitor, which will graphically display respiratory volumes. Note that respiratory volumes are measured in liters (l) along the Y-axis, and time in seconds is measured along the X-axis. Below the monitor are three buttons: **Normal Breathing, Hyperventilation,** and **Rebreathing.** Clicking any one of these buttons will induce the given pattern of breathing. Next to these buttons are three data displays for P_{CO_2} (partial pressure of carbon dioxide)—these will give us the levels of carbon dioxide in the blood over the course of an experimental run. At the very bottom of the screen is the data collection grid, where you may record and view your data after each activity.

Activity 1:
Normal Breathing

To get familiarized with the equipment, as well as to obtain baseline data for this experiment, we will first observe what happens during normal breathing.

1. Click **Start.** Notice that the **Normal Breathing** button dims, indicating that the simulated lungs are "breathing" normally. Also notice the reading in the pH meter at the top left, the readings in the P_{CO_2} displays, and the shape of the trace that starts running across the oscilloscope screen. As the trace runs, record the readings for pH at each of the following times:

At 20 seconds, pH = _____

At 40 seconds, pH = _____

At 60 seconds, pH = _____

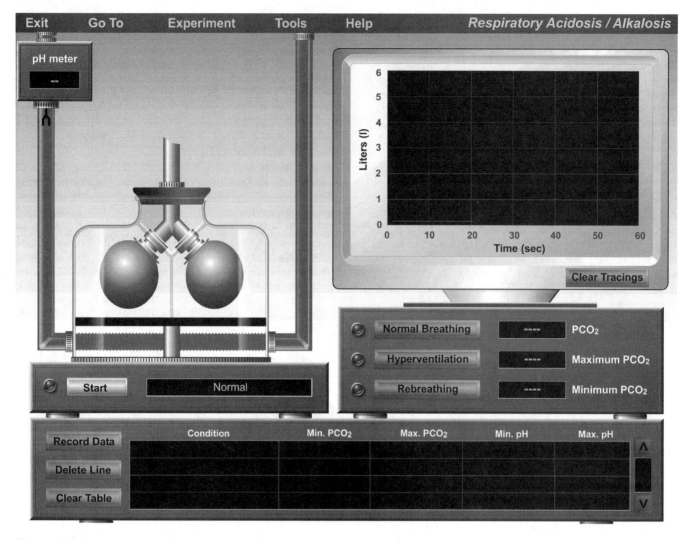

Figure 10.1 Opening screen of the Respiratory Acidosis/Alkalosis experiment.

2.   Allow the trace to run all the way to the right side of the oscilloscope screen. At this point, the run will automatically end.

3.   Click **Record Data** at the bottom left to record your results.

4.   If you have printer access, click **Tools** at the top of the screen and select **Print Graph.** Otherwise, manually sketch what you see on the oscilloscope screen.

5. Click **Clear Tracings** to clear the oscilloscope screen.

Did the pH level of the blood change at all during normal breathing? If so, how?

_____

_____

Was the pH level always within the "normal" range for the human body?

_____

_____

Did the P_{CO_2} level change during the course of normal breathing? If so, how?

_____

_____ ■

Activity 2a:
Hyperventilation—Run 1

Next, we will observe what happens to pH and carbon dioxide levels in the blood during hyperventilation.

1.   Click **Start.** Allow the normal breathing trace to run for 10 seconds; then at the 10-second mark, click **Hyperventilation.** Watch the pH meter display, as well as the readings in the P_{CO_2} displays and the shape of the trace. As the trace runs, record the readings for pH at each of the following times:

At 20 seconds, pH = _____

At 40 seconds, pH = _____

At 60 seconds, pH = _____

2. Allow the trace to run all the way across the oscilloscope screen and end.

3. Click **Record Data.**

4. If you have printer access, click **Tools** at the top of the screen and select **Print Graph.** Otherwise, manually sketch what you see on the oscilloscope screen on a separate sheet of paper.

5. Click **Clear Tracings** to clear the oscilloscope screen.

Did the pH level of the blood change at all during this run? If so, how?

_____

_____

Was the pH level always within the "normal" range for the

human body? _____

If not, when was the pH value outside of the normal range, and what acid-base imbalance did this pH value indicate?

_____

Did the P_{CO_2} level change during the course of this run? If so, how?

_____

If you observed an acid-base imbalance during this run, how would you expect the renal system to compensate for this condition?

_____

How did the hyperventilation trace differ from the trace for normal breathing? Did the tidal volumes change?

_____

_____

What might cause a person to hyperventilate?

_____ ∎

Activity 2b:
Hyperventilation—Run 2

This activity is a variation on Activity 2a.

1. Click **Start.** Allow the normal breathing trace to run for 10 seconds, then click **Hyperventilation** at the 10-second mark. Allow the hyperventilation trace to run for 10 seconds, then click **Normal Breathing** at the 20-second mark. Allow the trace to finish its run across the oscilloscope screen. Observe the changes in the pH meter and the P_{CO_2} displays.

2. Click **Record Data.**

3. If you have printer access, click **Tools** at the top of the screen and select **Print Graph.** Otherwise, manually sketch what you see on the oscilloscope screen.

4. Click **Clear Tracings** to clear the oscilloscope screen.

What happened to the trace after the 20-second mark when you stopped the hyperventilation? Did the breathing return to normal immediately? Explain your observation.

_____

_____

_____

_____ ∎

Activity 3:
Rebreathing

Rebreathing is the action of breathing in air that was just expelled from the lungs. Breathing into a paper bag is an example of rebreathing. In this activity, we will observe what happens to pH and carbon dioxide levels in the blood during rebreathing.

1. Click **Start.** Allow the normal breathing trace to run for 10 seconds; then at the 10 second mark, click **Rebreathing.** Watch the pH meter display, as well as the readings in the P_{CO_2} displays and the shape of the trace. As the trace runs, record the readings for pH at each of the following times:

At 20 seconds, pH = _____

At 40 seconds, pH = _____

At 60 seconds, pH = _____

2. Allow the trace to run all the way across the oscilloscope screen and end.

3. Click **Record Data.**

4. If you have printer access, click **Tools** at the top of the screen and select **Print Graph.** Otherwise, manually sketch what you see on the oscilloscope screen.

5. Click **Clear Tracings** to clear the oscilloscope screen.

Did the pH level of the blood change at all during this run? If so, how?

_____

Was the pH level always within the "normal" range for the

human body? _____

If not, when was the pH value outside of the normal range, and what acid-base imbalance did this pH value indicate?

_____

Did the P_{CO_2} level change during the course of this run? If so, how?

_____

_____

If you observed an acid-base imbalance during this run, how would you expect the renal system to compensate for this condition?

_____

_____

How did the rebreathing trace differ from the trace for normal breathing? Did the tidal volumes change?

_____

_____

_____

Give examples of respiratory problems that would result in pH and P_{CO_2} patterns similar to what you observed during rebreathing.

_____

_____

_____

6. To print out all of the recorded data from this activity, click **Tools** and then **Print Data.** ■

In the next set of activities, we will focus on the body's primary mechanism of compensating for respiratory acidosis or alkalosis: renal compensation.

Renal System Compensation

The kidneys play a major role in maintaining fluid and electrolyte balance in the body's internal environment. By regulating the amount of water lost in the urine, the kidneys defend the body against excessive hydration or dehydration. By regulating the excretion of individual ions, the kidneys maintain normal electrolyte patterns of body fluids. By regulating the acidity of urine and the rate of electrolyte excretion, the kidneys maintain plasma pH levels within normal limits. Renal compensation is the body's primary method of compensating for conditions of respiratory acidosis or respiratory alkalosis. (Although the renal system also compensates for metabolic acidosis or metabolic alkalosis, a more immediate mechanism for compensating for metabolic acid-base imbalances is the respiratory system, as we will see in a later experiment.)

The activities in this section examine how the renal system compensates for respiratory acidosis or alkalosis. The primary variable we will be working with is P_{CO_2} (the partial pressure of carbon dioxide in the blood). We will observe how increases and decreases in P_{CO_2} affect the levels of $[H^+]$ and $[HCO_3^-]$ (bicarbonate) that the kidneys excrete in urine.

Click on **Experiment** at the top of the screen and select **Renal System Compensation.** You will see the screen shown in Figure 10.2. If you completed Exercise 9 on renal

physiology, this screen should look familiar. There are two beakers on the left side of the screen, one of which is filled with blood, simulating the body's blood supply to the kidneys. Notice that the P_{CO_2} level is currently set to 40, and that the corresponding pH value is 7.4—both "normal" values. By clicking **Start,** you will initiate the process of delivering blood to the simulated nephron at the right side of the screen. As blood flows through the glomerulus of the nephron, you will see the filtration from the plasma of everything except proteins and cells (note that the moving red dots in the animation do *not* include red blood cells). Blood will then drain from the glomerulus to the beaker at the right of the original beaker. At the end of the nephron tube, you will see the collection of urine in a small beaker. Keep in mind that although only one nephron is depicted here, there are actually over a million nephrons in each human kidney. Below the urine beaker are displays for H^+ and HCO_3^-, which will tell us the relative levels of these ions present in the urine.

Activity 4:
Renal Response to Normal Acid-Base Balance

1. Set the P_{CO_2} value to 40, if it is not already. (To increase or decrease P_{CO_2}, click the (−) or (+) buttons. Notice that as P_{CO_2} changes, so does the blood pH level.)

2. Click **Start** and allow the run to finish.

3. At the end of the run, click **Record Data.**

At normal P_{CO_2} and pH levels, what level of H^+ was present

in the urine? _____

What level of $[HCO_3^-]$ was present in the urine? _____

Why does the blood pH value change as P_{CO_2} changes?

_____

_____

4. Click **Refill** to prepare for the next activity. ■

Activity 5:
Renal Response to Respiratory Alkalosis

In this activity, we will simulate respiratory alkalosis by setting the P_{CO_2} to values lower than normal (thus, blood pH will be *higher* than normal). We will then observe the renal system's response to these conditions.

1. Set P_{CO_2} to 35 by clicking the (−) button. Notice that the corresponding blood pH value is 7.5.

2. Click **Start.**

3. At the end of the run, click **Record Data.**

4. Click **Refill.**

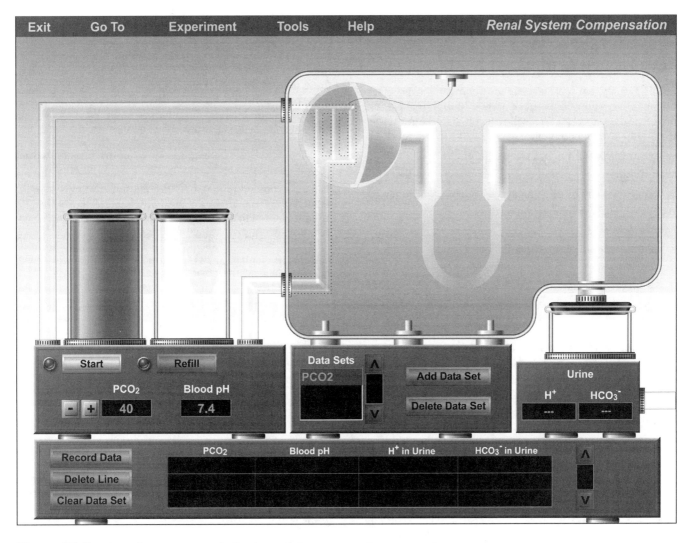

Figure 10.2 Opening screen of the Renal Compensation experiment.

5. Repeat steps 1–4, setting P_{CO_2} to increasingly lower values (i.e., set P_{CO_2} to 30 and then 20, the lowest value allowed).

What level of $[H^+]$ was present in the urine at each of these P_{CO_2}/pH levels?

_____

What level of $[HCO_3^-]$ was present in the urine at each of these P_{CO_2}/pH levels?

_____

Recall that it may take hours or even days for the renal system to respond to disruptions in acid-base balance. Assuming that enough time has passed for the renal system to fully compensate for respiratory alkalosis, would you expect P_{CO_2}

levels to increase or decrease? Would you expect blood pH levels to increase or decrease?

_____

_____

Recall your activities in the first experiment on respiratory acidosis and alkalosis. Which type of breathing resulted in P_{CO_2} levels closest to the ones we experimented with in this activity—normal breathing, hyperventilation, or rebreathing?

_____

_____

Explain why this type of breathing resulted in alkalosis.

_____

_____

Activity 6:
Renal Response to Respiratory Acidosis

In this activity, we will simulate respiratory acidosis by setting the P_{CO_2} values higher than normal (thus, blood pH will be *lower* than normal). We will then observe the renal system's response to these conditions.

1. Make sure the left beaker is filled with blood. If not, click **Refill**.

2. Set P_{CO_2} to 60 by clicking the (+) button. Notice that the corresponding blood pH value is 7.3.

3. Click **Start**.

4. At the end of the run, click **Record Data**.

5. Click **Refill**.

6. Repeat steps 1–5, setting P_{CO_2} to increasingly higher values (i.e., set P_{CO_2} to 75 and then 90, the highest value allowed).

What level of $[H^+]$ was present in the urine at each of these P_{CO_2}/pH levels?

_____

_____

What level of $[HCO_3^-]$ was present in the urine at each of these P_{CO_2}/pH levels?

_____

_____

Recall that it may take hours or even days for the renal system to respond to disruptions in acid-base balance. Assuming that enough time has passed for the renal system to fully compensate for respiratory acidosis, would you expect P_{CO_2} levels to increase or decrease? Would you expect blood pH levels to increase or decrease?

_____

_____

Recall your activities in the first experiment on respiratory acidosis and alkalosis. Which type of breathing resulted in P_{CO_2} levels closest to the ones we experimented with in this activity—normal breathing, hyperventilation, or rebreathing?

_____

Explain why this type of breathing resulted in acidosis.

_____

_____

7. Before going on to the next activity, select **Tools** and then **Print Data** in order to save a hard copy of your data results. ■

Metabolic Acidosis and Alkalosis

Conditions of acidosis or alkalosis that do not have respiratory causes are termed *metabolic acidosis* or *metabolic alkalosis*.

Metabolic acidosis is characterized by low plasma HCO_3^- and pH. The causes of metabolic acidosis include:

- *Ketoacidosis,* a buildup of keto acids that can result from diabetes mellitus

- *Salicylate poisoning,* a toxic condition resulting from ingestion of too much aspirin or oil of wintergreen (a substance often found in laboratories)

- The ingestion of too much alcohol, which metabolizes to acetic acid

- Diarrhea, which results in the loss of bicarbonate with the elimination of intestinal contents

- Strenuous exercise, which may cause a buildup of lactic acid from anaerobic muscle metabolism

Metabolic alkalosis is characterized by elevated plasma HCO_3^- and pH. The causes of metabolic alkalosis include:

- Alkali ingestion, such as antacids or bicarbonate

- Vomiting, which may result in the loss of too much H^+

- Constipation, which may result in reabsorption of elevated levels of HCO_3^-

Increases or decreases in the body's normal metabolic rate may also result in metabolic acidosis or alkalosis. Recall that carbon dioxide—a waste product of metabolism—mixes with water in plasma to form carbonic acid, which in turn forms H^+:

$$H_2O + \mathbf{CO_2} \rightarrow \underset{\substack{\text{carbonic} \\ \text{acid}}}{H_2CO_3} \rightarrow \mathbf{H^+} + \underset{\substack{\text{bicarbonate} \\ \text{ion}}}{HCO_3^-}$$

Therefore, an increase in the normal rate of metabolism would result in more carbon dioxide being formed as a metabolic waste product, resulting in the formation of more H^+—lowering plasma pH and potentially causing acidosis. Other acids that are also normal metabolic waste products, such as ketone bodies and phosphoric, uric, and lactic acids, would likewise accumulate with an increase in metabolic rate. Conversely, a decrease in the normal rate of metabolism would result in less carbon dioxide being formed as a metabolic waste product, resulting in the formation of less H^+—raising plasma pH and potentially causing alkalosis. Many factors can affect the rate of cell metabolism. For example, fever, stress, or the ingestion of food all cause the rate of cell metabolism to increase. Conversely, a fall in body temperature or a decrease in food intake causes the rate of cell metabolism to decrease.

The respiratory system compensates for metabolic acidosis or alkalosis by expelling or retaining carbon dioxide in the blood. During metabolic acidosis, respiration increases to expel carbon dioxide from the blood and decrease $[H^+]$ in order to raise the pH level. During metabolic alkalosis, res-

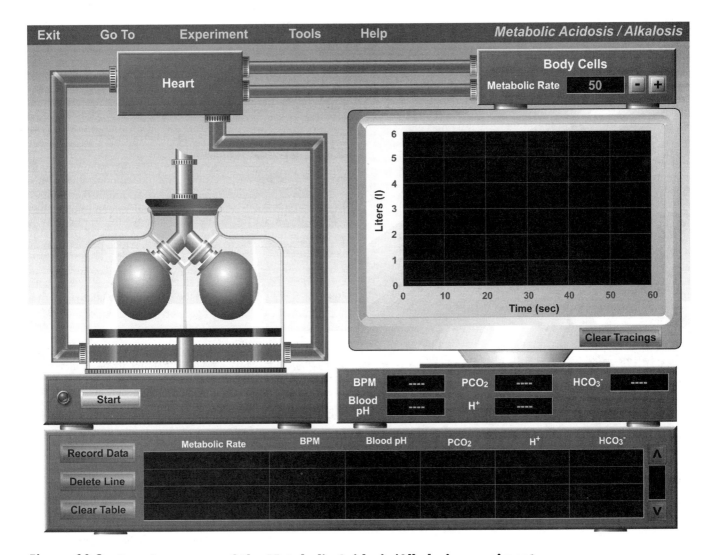

Figure 10.3 Opening screen of the Metabolic Acidosis/Alkalosis experiment.

piration decreases to promote the accumulation of carbon dioxide in the blood, thus increasing $[H^+]$ and decreasing the pH level.

The renal system also compensates for metabolic acidosis and alkalosis by conserving or excreting bicarbonate ions. However, in this set of activities we will focus on respiratory compensation of metabolic acidosis and alkalosis.

To begin, click **Experiment** at the top of the screen and select **Metabolic Acidosis/Alkalosis.** The screen shown in Figure 10.3 will appear. This screen is similar to the screen from the first experiment; the main differences are the addition of a box representing the heart; tubes showing the double circulation of the heart; and a box representing the body's cells. The default "normal" metabolic rate has been set to 50 kcal/h—an arbitrary value, given that "normal" metabolic rates vary widely from individual to individual. The (+) and (−) buttons in the Body Cells box allow you to increase or decrease the body's metabolic rate. In the following activities, we will observe the respiratory response to acidosis or alkalosis brought on by increases or decreases in the body's metabolic rate.

Activity 7:

Respiratory Response to Normal Metabolism

We will begin by observing respiratory activity at normal metabolic conditions. This data will serve as a baseline against which we will compare our data in Activities 8 and 9.

1. Make sure the Metabolic Rate is set to 50, which for the purposes of this experiment we will consider the "normal" value.

2. Click **Start** to begin the experiment. Notice the arrows showing the direction of blood flow. A graph displaying respiratory activity will appear on the oscilloscope screen.

3. After the graph has reached the end of the screen, the experiment will automatically stop. Note the data in the displays below the oscilloscope screen:

• The **BPM** display gives you the *breaths-per-minute—* the rate at which respiration occurred.

- Blood pH tells you the pH value of the blood.
- PCO_2 (shown as P_{CO_2} in the text) tells you the partial pressure of carbon dioxide in the blood.
- H^+ and HCO_3^- tell you the levels of each of these ions present.

4.   Click **Record Data**.

5.   Click **Tools** and then **Print Graph** in order to print your graph.

What is the respiratory rate? _____

Are the blood pH and P_{CO_2} values within normal ranges?

_____

6.   Click **Clear Tracings** before proceeding to the next activity. ■

Activity 8:
Respiratory Response to Increased Metabolism

1.   Increase the metabolic rate to 60.

2.   Click **Start** to begin the experiment.

3.   Allow the graph to reach the end of the oscilloscope screen. Note the data in the displays below the oscilloscope screen.

4.   Click **Record Data**.

5.   Click **Tools** and then **Print Graph** in order to print your graph.

6.   Repeat steps 1–5 with the metabolic rate set at 70, and then 80.

As the body's metabolic rate increased:

How did respiration change?

_____

How did blood pH change?

_____

How did P_{CO_2} change?

_____

How did $[H^+]$ change?

_____

How did $[HCO_3^-]$ change?

_____

Explain why these changes took place as metabolic rate increased.

_____

_____

Which metabolic rates caused pH levels to decrease to a condition of metabolic acidosis?

_____

What were the pH values at each of these rates?

_____

By the time the respiratory system fully compensated for acidosis, how would you expect the pH values to change?

_____

7.   Click **Clear Tracings** before proceeding to the next activity. ■

Activity 9:
Respiratory Response to Decreased Metabolism

1.   Decrease the metabolic rate to 40.

2.   Click **Start** to begin the experiment.

3.   Allow the graph to reach the end of the oscilloscope screen. Note the data in the displays below the oscilloscope screen.

4.   Click **Record Data**.

5.   Click **Tools** and then **Print Graph** in order to print your graph.

6.   Repeat steps 1–5 with the metabolic rate set at 30, and then 20.

As the body's metabolic rate decreased:

How did respiration change?

decrease

How did blood pH change?

increase

How did P_{CO_2} change?

decrease

How did $[H^+]$ change?

decrease

How did [HCO$_3$$^-$] change?

_____

Explain why these changes took place as the metabolic rate decreased.

_____

_____

_____

Which metabolic rates caused pH levels to increase to a condition of metabolic alkalosis?

_____

What were the pH values at each of these rates?

_____

By the time the respiratory system fully compensated for acidosis, how would you expect the pH values to change?

_____

7. Click **Tools** → **Print Data** to print your recorded data. ■

Using the Histology Tutorial

Objectives

1. To understand that organ function is the result of cell function
2. To understand that cell function is closely associated with cell structure
3. To understand that magnification and resolution are two different things
4. To learn how to take your own histology photos

To the physiology student, **histology**—the study of tissues—may seem inappropriate at first. However, organs are groups of tissues, and tissues are groups of similar cells that perform similar functions. By studying tissues under the microscope and learning what the organelles of a cell do, we can obtain a better understanding of cell—and subsequently organ—physiology.

A common mistake students make when studying tissues under a microscope is mixing up the terms **magnification** and **resolution.** Magnification, by itself, just means that what is being looked at is now larger. Magnification has two drawbacks. First, magnification does little for actually seeing anything. What is important is *resolution*—how close two points can be and still be recognized as two distinct points. What makes electron microscopy so important to biologists is that the electron microscope allows for such magnificent resolution. The second problem with magnification is that when one switches to an increased magnification, the field of vision is reduced. Therefore, everything visible under the lower magnification may not be in view at the higher power magnification.

To access the histology tutorial, select **Histology Tutorial** from the Main Menu. You will see the opening screen shown in Figure H.1 on page 100. To select an image, click **Select an Image,** scroll through the alphabetical index, and click on the slide you wish to view. The slide will appear in the large monitor on the right side of the screen, with accompanying text to the left. If you see any words in the accompanying text that appear in purple (or hot pink, depending on your monitor), click them—they link to another slide that will appear in the small window at the lower left corner of your screen. Click the **Close** button next to this small window in order to close itbefore selecting a new slide to view. You may change slide magnifications by clicking the $\times$ 40, $\times$ 100, $\times$ 400, or $\times$ 1000 buttons (although note that when these buttons are red, this indicates that the slide is not available at that magnification.)

By clicking your mouse on an image and moving the mouse around (without releasing the mouse button), you can move the slide around—much like moving a slide around under a microscope. You may also see the slide with its components labeled by clicking the **Labels On** button in the lower right corner of the screen. Click the button again to remove the labels.

All tissues have an epithelium, or layer of covering cells. All epithelial tissues are found on a "basement membrane," or *basal lamina.* Usually below this is a layer of connective tissue that helps hold the organ or tissue together. In some organs, such as the intestine, layers of muscle are found below the connective tissue and are responsible for the intestinal function of peristalsis. To best understand histology in relation to physiology, take notes regarding the epithelium of a given tissue, and which layers are found below

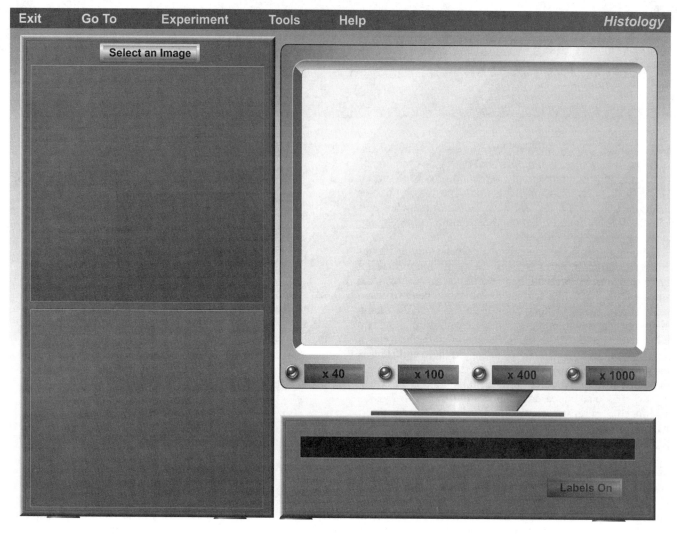

Figure H.1   Opening screen of the Histology experiment.

the epithelium; then look for these when you are examining the tissue slide. Do not pay attention to tissue staining—what is important is the arrangement of the tissue layers.

Histology slides like the ones in this module can be taken by any camera, including digital cameras. If you have access to a microscope, turn on the light source. Take a piece of white paper and stretch it over the eyepiece. As you move the paper from the eyepiece up, you will see a spot of light on the paper from the microscope. Move the paper up and down until you get the smallest point of light. Place your camera lens at this point and be sure that the film plane is perpendicular to

the light plane. A tripod can be useful for this. You can take the picture directly. If your camera has a self-timer, use it so that there is no movement of the camera while taking the picture. Taking a picture this way will result in a circular image, which can be easily adjusted to a rectangular trim if you are using a digital camera and have photo editing software such as *Adobe PhotoShop*.

For a review of histology slides specific to topics covered in the PhysioEx lab simulations, turn to the Histology Review Supplement on p. P-135.

Cell Transport Mechanisms and Permeability

1. Match each of the definitions in Column A with the appropriate term in Column B.

Column A

_____ term used to describe a solution that has a *lower* concentration of solutes compared to another solution

_____ term used to describe a solution that has a *higher* concentration of solutes compared to another solution

_____ the movement of molecules from an area of higher concentration to an area of lower concentration as a result of random thermal motion

_____ the movement of molecules across a membrane that requires the expenditure of cellular energy (ATP)

_____ the transport of water across a semipermeable membrane

_____ term used to describe two solutions that have the same concentration of solutes relative to one another

_____ the movement of molecules across a selectively permeable membrane with the aid of specialized transport proteins

Column B

a. diffusion

b. facilitated diffusion

c. osmosis

d. active transport

e. hypotonic

f. isotonic

g. hypertonic

2. What is the main difference between simple diffusion and facilitated diffusion?

_____

_____

_____

3. What is the main difference between facilitated diffusion and active transport?

_____

_____

_____

4. In the "Simple Diffusion" experiment, which solute(s) passed through the MWCO 20 membrane?

_____

Why?

_____

5. List three examples of passive transport mechanisms.

_____

_____

_____

6. Describe the relationship of solute concentration to solvent concentration in osmosis.

_____

_____

_____

7. What is the equation for Fick's First Law of Diffusion?

_____

Explain Fick's First Law of Diffusion.

_____

_____

_____

8. In the mock dialysis activity, what was the only solute removed from the beaker representing the patient's blood?

_____

Why is it important that this solute be removed from diabetic patients?

_____

_____

_____

9. How can the concentration of water in a solution be decreased?

_____

_____

_____

10. Suppose that a membrane separates a solution of higher osmolarity and a solution of lower osmolarity. To prevent osmotic flow of water across the membrane, pressure should be applied to which of the two solutions?

_____

11. What change in cell volume will occur when a cell is placed in a hypotonic solution?

_____

_____

_____

12. What change in cell volume will occur when a cell is placed in a hypertonic solution?

_____

_____

_____

13. By what mechanism does the active transport of sodium lead to osmotic flow of water across a membrane?

_____

_____

_____

_____

_____

_____

14. If two solutions having different osmolarities are separated by a water-permeable membrane, will there be a change in the volume of the two compartments if the membrane is impermeable to solutes?

_____

Will there be a change in the volume of the two compartments if the membrane is permeable to solutes?

_____

Explain your answers.

_____

_____

_____

Skeletal Muscle Physiology

1. Define each of the following terms:

- motor unit _____

- twitch _____

- threshold _____

- treppe _____

- summation _____

- tetanus _____

- fatigue _____

- isometric contraction _____

- isotonic contraction _____

2. Describe the process of excitation-contraction coupling.

_____

 • _____

_____

_____

_____

3. What is the role of acetylcholine in a muscle contraction?

_____

_____

_____

_____

4. Describe the three phases of a muscle "twitch."

_____

_____

_____

_____

_____

5. What could be a chemical cause of fatigue?

_____

6. In fatigue, what happens to force production over time?

_____

7. If you were lifting a dumbbell, would your muscles be contracting isometrically or isotonically?

_____

_____

_____

_____

8. What is the key variable in an isometric contraction?

_____

9. Define the term _maximal stimulus._

_____

10. What has happened in the muscle when the maximal stimulus is achieved?

_____

_____

11. What is the difference between stimulus intensity and stimulus frequency?

_____

_____

_____

12. Circle the correct boldfaced term.

At the threshold stimulus, sodium ions start to move **into / out of** the cell to bring about the membrane depolarization.

Neurophysiology of Nerve Impulses

1. Match each of the definitions in Column A with the appropriate term in Column B.

Column A

_____ term that refers to a membrane potential of about −70 mv

_____ reversal of membrane potential due to influx of sodium ions

_____ major cation found outside of a cell

_____ minimal stimulus needed to elicit an action potential

_____ period when cell membrane is totally insensitive to additional stimuli, regardless of the stimulus force applied

_____ major cation found inside of a cell

Column B

a.   threshold

b.   sodium

c.   potassium

d.   resting membrane potential

e.   absolute refractory period

f.   depolarization

2. Fill in the blanks with the correct words or terms.

Neurons, as with other excitable cells of the body, have two major physiological properties: _____ and

_____. A neuron has a positive charge on the outer surface of the cell membrane due in part to the action of

an active transport system called the _____. This system moves _____

out of the cell and _____ into the cell. The inside of the cell membrane is negative, not only due to the active

transport system but also because of _____, which remain negative due to intracellular pH and keep the

inside of the cell membrane negative.

3. Why don't the terms *depolarization* and *action potential* mean the same thing?

_____

_____

_____

4. What is the difference between membrane irritability and membrane conductivity?

_____

_____

_____

5. Why does a nerve's action potential increase slightly when you add 1.0 V to the threshold voltage and stimulate the nerve?

_____

_____

_____

6. If you were to spend a lot of time studying nerve physiology in the laboratory, what type of stimulus would you use, and why?

_____

_____

_____

7. Why does the addition of sodium chloride elicit an action potential?

_____

_____

8. What was the effect of ether on eliciting an action potential?

_____

9. Does the addition of ether to the nerve cause any permanent alteration in neural response?

_____

10. What was the effect of curare on eliciting an action potential?

_____

11. Explain the reason for your answer to Question 10.

_____

12. What was the effect of lidocaine on eliciting an action potential?

_____

13. What is the relationship between size of a nerve and conduction velocity?

_____

_____

14. Keeping your answer to Question 13 in mind, draw an analogy between the nerves in the human body and electrical wires.

_____

15. Hypothesize what types of animals would have the fastest conduction velocities.

_____

16. How does myelination affect nerve conduction velocity? Explain.

_____

_____

_____

17. In the nerve conduction velocity experiment, if any of the nerves used were reversed in their placement on the stimulating and recording electrodes, would there be any differences seen in conduction velocity? Explain.

_____

_____

_____

Endocrine System Physiology

1. Match each of the hormones in the left hand column with its source.

_____ thyroxine

_____ estrogen

_____ thyroid stimulating hormone (TSH)

_____ insulin

   a. ovary

   b. thyroid gland

   c. pancreas

   d. pituitary gland

2. Each hormone is known to have a specific target tissue. For each of the following hormones, list its target tissue and describe its specific action.

thyroxine   _____

estrogen   _____

    _____

thyroid stimulating hormone (TSH)   _____

    _____

insulin   _____

    _____

follicle stimulating hormone (FSH)   _____

    _____

3. What is the role of the hypothalamus in the production of thyroxine and TSH?

_____

_____

4. How does thyrotropin releasing hormone (TRH) travel from the hypothalamus to the pituitary gland?

_____

_____

5. What are *tropic* hormones?

_____

_____

6. In the metabolism experiment, what was the effect of thyroxine on the overall metabolic rate of the animals?

_____

_____

7. Using the respirometer-manometer, you observed the amount of oxygen being used by animals in a closed chamber. What happened to the carbon dioxide the animals produced while in the chamber?

_____

_____

8. (a) If the experimental animals in the chamber were engaged in physical activity (such as running in a wheel), how would this change the results of the metabolism experiment?

_____

_____

(b) What changes would you expect to see in fluid levels of the manometer?

_____

_____

9. Why didn't the administration of thyroid stimulating hormone (TSH) have any effect on the metabolic rate of the thyroidectomized rat?

_____

_____

10. Why didn't the administration of propylthiouracil have any effect on the metabolic rate of either the thyroidectomized rat or the hypophysectomized rat?

_____

_____

_____

_____

11. In the hormone replacement therapy experiment, what was the effect of removing the ovaries from the animals?

_____

12. Specifically, what hormone did the ovariectomies effectively remove from the animals, and what purpose does this hormone serve?

_____

_____

13. If a hormone such as testosterone were used in place of estrogen, would any effect be seen? Explain your answer.

_____

_____

14. In the experiment, you administered seven injections of estrogen to the experimental rat over the course of 7 days. What do you think would happen if you administered one injection of estrogen per day for an additional week?

_____

_____

15. What do you think would happen if you administered seven injections of estrogen to the experimental rat all in one day?

_____

_____

16. In a wet lab, why would you need to wait several weeks after the animals underwent their ovariectomies before you could perform this experiment on them?

_____

_____

17. In the insulin and diabetes experiment, what was the effect of administering alloxan to the experimental animal?

_____

18. (a) When insulin travels to the cells of the body, the concentration of what compound will elevate within the cells?

_____

(b) What is the specific action of this compound, within the cells?

_____

_____

19. Fill in the blanks:

(a) The condition when insulin is not produced by the pancreas:

_____

(b) The condition when insulin is produced by the pancreas, but the body fails to respond to the insulin:

_____

20. What was the effect of administering insulin to the diabetic rat?

_____

_____

21. What is a glucose standard curve, and why did you need to obtain one for this experiment?

_____

_____

_____

22. Would altering the light wavelength of the spectrophotometer have any bearing on the results obtained? Explain your answer.

_____

_____

23. What would you do to help a friend who had inadvertently taken an overdose of insulin? Why?

_____

_____

Cardiovascular Dynamics

1. Identify each of the following variables in Poiseuille's equation:

    ΔP = _____

    r^4 = _____

    η = _____

    l = _____

2. Explain how each of the following variables affects blood flow.

    ΔP: _____

    _____

    r^4: _____

    _____

    η: _____

    _____

    _____

    l: _____

    _____

3. What could cause an increase in the peripheral resistance in a blood vessel?

    _____

    _____

    _____

4. Describe the cardiac cycle.

    _____

    _____

    _____

    _____

5. Match each of the definitions in Column A with the appropriate term in Column B.

Column A

_____ ventricular contraction

_____ the amount of blood each ventricle pumps per minute

_____ the amount of blood pumped to the body per contraction per ventricle

_____ the volume of blood in the heart at the end of ventricular contraction

_____ ventricular relaxation

_____ contraction that occurs when the volume of blood in the ventricles remains constant

_____ the ejection of blood near the end of systole during which ventricular pressure rises and then begins to decline

_____ the volume of blood in the heart at the end of ventricular relaxation

Column B

a. diastole

b. systole

c. end diastolic volume (EDV)

d. end systolic volume (ESV)

e. cardiac output

f. stroke volume

g. isovolumetric contraction

h. ventricular ejection

6. Define Starling's Law.

_____

_____

_____

7. What differences would you expect to see between a diseased heart with high peripheral resistance and the healthy heart of an athlete?

_____

_____

_____

8. What was the effect of increasing flow tube radius on flow rate and flow volume?

_____

_____

9. Which variable had the strongest effect on fluid flow?

_____

_____

10. If the viscosity of blood were to increase, what could you do to keep the flow rate "normal"?

_____

_____

11. What would occur if the left side of the heart pumped faster than the right side?

_____

_____

12. What do the valves in the *Pump Mechanics* screen do?

_____

_____

13. Match the terms in the right hand column to the simulation equipment from the *Pump Mechanics* screen.

_____ fluid in left beaker   a.   vein

_____ middle beaker   b.   blood going to the rest of the body

_____ fluid in right beaker   c.   artery

_____ flow tube between left and middle beakers   d.   blood coming from the lungs

_____ flow tube between middle and right beakers   e.   left side of the heart

Cardiovascular Physiology

1. Define each of the following terms:

   • autorhymicity _____

   • sinoatrial node _____

   • pacemaker cells _____

   • vagus nerves _____

2. The sympathetic nervous system releases the neurotransmitter _____ .

3. The parasympathetic nervous system releases the neurotransmitter _____ .

4. Circle the correct boldfaced term.

   The sympathetic nervous system **increases / decreases** heart rate.

   The parasympathetic nervous system **increases / decreases** heart rate.

5. What happens in each of the five phases of cardiac muscle depolarization?

   Phase 0: _____

   _____

   Phase 1: _____

   _____

   Phase 2: _____

   _____

   Phase 3: _____

   _____

   Phase 4: _____

   _____

6. Explain why the SA node generates action potentials at a frequency of approximately 100 beats per minute even though the average resting heart rate is 70 beats per minute.

   _____

   _____

7. What are two key differences between cardiac muscle and skeletal muscle?

_____

_____

_____

_____

8. What is the difference between the effective refractory period and the relative refractory period?

_____

_____

_____

_____

9. When the heart is externally stimulated just after the start of the contraction cycle, why does this have no affect on heart rate?

_____

_____

10. What is the action of each of the following factors on heart rate?

epinephrine: _____

pilocarpine: _____

atropine: _____

digitalis: _____

temperature: _____

11. How do a frog heart and a human heart differ in their responses to temperature? Why?

_____

12. What is the action of each of the following ions on heart rate?

Calcium: _____

Sodium: _____

Potassium: _____

Respiratory System Mechanics

1. Define each of the following terms:

   • respiration _____

   • ventilation _____

   • alveoli _____

   • diaphragm _____

   • inspiration _____

   • expiration _____

2. Explain how the respiratory and circulatory systems work together to distribute oxygen to, and remove carbon dioxide from, the cells of the body.

   _____

   _____

3. Match each of the definitions in Column A with the appropriate term in Column B.

Column A

_____ the percentage of vital capacity exhaled during a 1-second period of the FVC test

_____ the amount of air that can be taken into the lungs beyond the tidal volume

_____ the amount of air that can be expelled from the lungs beyond the tidal volume

_____ the volume of a normal breath

_____ the maximum amount of air that can be voluntarily moved in and out of the lungs

_____ the proportion of pressure that a single gas exerts within a mixture

_____ the amount of air that can be expelled completely and rapidly as possible after a maximum inspiration

_____ the amount of air left in the lungs after a maximum exhalation

_____ vital capacity plus residual volume

Column B

a. tidal volume

b. partial pressure

c. inspiratory reserve volume

d. expiratory reserve volume

e. vital capacity

f. residual volume

g. total lung capacity

h. forced vital capacity (FVC)

i. forced expiratory volume (FEV$_1$)

4. Fill in the typical values (in ml) for each of the following terms.

tidal volume: _____

inspiratory reserve volume: _____

expiratory reserve volume: _____

vital capacity: _____

residual volume: _____

total lung capacity: _____

5. Circle the correct boldfaced term.

Emphysema is a lung problem that causes a(n) **decrease / increase** in tidal volume.

6. How do you calculate minute respiratory volume?

_____

7. What was the effect of reducing the radius of the air flow tube on respiratory volumes?

_____

8. What is the role of surfactant in respiration?

_____

_____

9. What would happen if surfactant were not present?

_____

_____

10. What happens in pneumothorax?

_____

11. Why is it important that intrathoracic pressure be kept lower than atmospheric pressure?

_____

_____

_____

12. What happens to the partial pressure of carbon dioxide in the blood during rapid breathing?

_____

13. What happens to the partial pressure of carbon dioxide during rebreathing?

_____

14. What happens to the partial pressure of carbon dioxide during breathholding?

_____

_____

Chemical and Physical Processes of Digestion

1. Define each of the following terms:

   • digestive tract _____

   • accessory glands _____

   • digestion _____

   • hydrolase _____

   • salivary amylase _____

   • bile salts _____

   • pepsin _____

   • lipase _____

2. List two factors that play key roles in the efficacy of digestive enzymes, and explain their effects.

   _____

   _____

   _____

   _____

3. What are the major functions of the digestive system?

   _____

   _____

4. What does it mean for an enzyme to be substrate specific?

   _____

   _____

5. What are the three primary categories into which food molecules fall?

   _____

   _____

6. Complete the following sentences.

Carbohydrates are broken down into _____

Proteins are broken down into _____

Lipids are broken down into _____

7. Why do lipids pose special problems for digestion?

_____

_____

_____

8. What is meant by hydrolysis?

_____

_____

_____

9. Match each digestive enzyme with the type of food molecules upon which it acts.

_____ carbohydrates   a. amylase

_____ lipids   b. pepsin

_____ proteins   c. lipase

10. Why was 37°C the appropriate temperature to use in the experiment on salivary amylase?

_____

11. What was the optimal pH for salivary amylase? _____

12. Compare the effects of boiling and freezing on enzyme activity.

_____

_____

13. Was salivary amylase able to digest cellulose? _____

14. What was the effect of bacteria on cellulose digestion?

_____

15. Where is pepsin secreted? _____

16. What was the optimal pH for pepsin? _____

17. Where is lipase secreted? _____

18. What was the optimal pH for lipase? _____

19. What was the effect of bile salts on lipid digestion?

_____

_____

_____

_____

20. Explain what each of the following was used to test:

IKI: _____

Benedict's: _____

spectrophotometer: _____

pH meter: _____

21. Give some examples of physical processes of digestion.

_____

_____

_____

Renal System Physiology

1. Define each of the following terms:

   • nephron _____

   • renal corpuscle _____

   • renal tubule _____

   • afferent arteriole _____

   • glomerular filtration _____

   • efferent arteriole _____

   • aldosterone _____

   • ADH _____

   • reabsorption _____

2. What are the primary functions of the kidneys?

   _____

   _____

   _____

   _____

3. What are the components of the renal corpuscle?

   _____

   _____

4. What are the parts of the renal tubule?

   _____

   _____

5. What drives protein-free plasma from the blood into the Bowman's capsule?

   _____

6. How can the glomerular filtration rate be changed?

_____

_____

7. Is most filtrate reabsorbed into the body or excreted in urine? Explain.

_____

_____

8. What does the hormone aldosterone regulate? _____

_____

9. What does the hormone ADH regulate? _____

_____

10. What is the effect of reducing afferent arteriole radius on filtration rate?

_____

11. What is the effect of reducing efferent arteriole radius on filtration rate?

_____

12. What effects would increased blood pressure have on nephron function?

_____

_____

_____

13. What could cause increased blood pressure in the glomerulus?

_____

_____

14. How could you adjust the afferent or efferent radius to compensate for the effect of reduced blood pressure on glomerular filtration rate?

_____

15. What common condition is analogous to the valve above the collecting duct being shut off?

_____

16. Once solutes have been filtered from the plasma into the filtrate, how are they reabsorbed into the blood?

_____

_____

_____

_____

17. Can the reabsorption of solutes influence water reabsorption in the nephron? Explain.

_____

_____

18. What happens as the concentration gradient of the interstitial fluid increases?

_____

_____

19. What type of transport is used during glucose reabsorption?

_____

20. How could glucose carrier proteins in the kidney become overwhelmed?

_____

21. What happens to urine volume when water intake is decreased?

_____

22. What happens to urine volume when water intake is increased?

_____

23. How does ADH affect the concentration of potassium in urine?

_____

24. How does aldosterone affect the concentration of potassium in urine?

_____

25. If ADH were not present, what would be the effect on urine concentration?

_____

26. What is the principal determinant for the release of aldosterone?

_____

27. What is the principal determinant for the release of ADH?

_____

28. What would cause the body to excrete more sodium?

_____

29. What is the major change seen when both ADH and aldosterone are added to the system? Why?

_____

_____

Acid-Base Balance

1. Match each of the terms in column A with the appropriate description in column B.

Column A

_____ 1. pH

_____ 2. acid

_____ 3. base

_____ 4. acidosis

_____ 5. alkalosis

_____ 6. carbon dioxide

Column B

a. condition in which the human body's pH levels fall below 7.35

b. condition in which the human body's pH levels rise above 7.45

c. mixes with water in the blood to form carbonic acid

d. substance that binds to H^+ in solution

e. substance that releases H^+ in solution

f. term used to denote hydrogen ion concentration in body fluids

2. What is the normal range of pH levels of blood and tissue fluids in the human body?

_____

3. What is the difference between a *strong acid* and a *weak acid*?

_____

4. What is the difference between a *strong base* and a *weak base*?

_____

5. What is the difference between respiratory acidosis/alkalosis and metabolic acidosis/alkalosis?

_____

_____

6. What are the body's two major physiological buffer systems for compensating for acid-base imbalances?

_____

Respiratory Acidosis and Alkalosis

1. What are some of the causes of respiratory acidosis?

_____

2. What are some of the causes of respiratory alkalosis?

_____

3. What happens to blood pH levels during hyperventilation? Why?

_____

4. What happens to blood pH levels during rebreathing? Why?

_____

_____

5. Circle the correct bolfaced terms:

As respiration increases, P_{CO_2} levels **increase / decrease** and pH levels **rise / fall.**

As respiration decreases, P_{CO_2} levels **increase / decrease** and pH levels **rise / fall.**

Renal Compensation

1. How does the renal system compensate for conditions of respiratory acidosis?

_____

2. How does the renal system compensate for conditions of respiratory alkalosis?

_____

Metabolic Acidosis and Alkalosis

1. What are some of the causes of metabolic acidosis?

_____

2. What are some of the causes of metabolic alkalosis?

_____

3. Explain how the respiratory system compensates for metabolic acidosis and alkalosis.

_____

_____

_____

4. Explain how the renal system compensates for metabolic acidosis and alkalosis.

_____

_____

_____

5. Circle the correct bolfaced terms:

As metabolic rate increases, respiration **increases / decreases,** P_{CO_2} levels **increase / decrease,** and pH levels **rise / fall.**

As metabolic rate decreases, respiration **increases / decreases,** P_{CO_2} levels **increase / decrease,** and pH levels **rise / fall.**

Histology Review Supplement

The slides in this section are designed to provide a basic histology review related to topics introduced in the PhysioEx lab simulations and in your anatomy and physiology textbook.

From the PhysioEx main menu, select **Histology Review Supplement.** When the screen comes up, click **Select an Image Group.** You will note that the slides in the histology module are grouped in the following categories:

Skeletal muscle slides

Nervous tissue slides

Endocrine tissue slides

Cardiovascular tissue slides

Respiratory tissue slides

Digestive tissue slides

Renal tissue slides

Select the group of slides you wish to view, and then refer to the relevant worksheet in this section for a step-by-step tutorial. For example, if you would like to review the skeletal muscle slides, click on **Skeletal muscle slides** and then turn to the next page of this lab manual for the worksheet entitled Skeletal Muscle Tissue Review to begin your review.

Since the slides in this module have been selected for their relevance to topics covered in the PhysioEx lab simulations, it is recommended that you complete the worksheets along with a related PhysioEx lab. For example, you might complete the Skeletal Muscle Tissue worksheet right before or after your instructor assigns you Exercise 2, the PhysioEx lab simulation on Skeletal Muscle Physiology.

For additional histology review, turn to page 99.

Skeletal Muscle Tissue Review

From the PhysioEx main menu, select **Histology Review Supplement.** When the screen comes up, click **Select an Image Group**. From Group Listing, click **Skeletal Muscle Slides**. To view slides without labels, click the **Labels Off** button at the bottom right of the monitor.

Click slide 1.
Skeletal muscle is composed of extremely large, cylindrical multinucleated cells called **myofibers.** The nuclei of the skeletal muscle cell (**myonuclei**) are located peripherally just subjacent to the muscle cell plasmalemma (sarcolemma). The interior of the cell is literally filled with an assembly of contractile proteins (myofilaments) arranged in a specific overlapping pattern oriented parallel to the long axis of the cell.

Click slides 2, 3.
Sarcomeres are the functional units of skeletal muscle. The organization of contractile proteins into a regular end-to-end repeating pattern of sarcomeres along the length of each cell accounts for the striated or striped appearance of skeletal muscle in longitudinal section.

Click slide 4.
The smooth endoplasmic reticulum (sarcoplasmic reticulum), modified into an extensive network of membranous channels that store, release, and take up the calcium necessary for contraction, also functions to further organize the myofilaments inside the cell into cylindrical bundles called myofibrils. The **stippled appearance of the cytoplasm** in cells cut in cross section represents the internal organization of myofilaments bundled into myofibrils by the membranous sarcoplasmic reticulum.

What is the functional unit of contraction in skeletal muscle?

_____

What are the two principal contractile proteins that compose the functional unit of contraction?

_____

What is the specific relationship of the functional unit of contraction to the striated appearance of a skeletal muscle fiber?

_____

_____

Click slide 5.
The neural stimulus for contraction arises from the **axon** of a motor neuron whose axon terminal comes into close apposition to the muscle cell sarcolemma.

Would you characterize skeletal muscle as voluntary or involuntary?

_____

Name the site of close juxtaposition of an axon terminal with the muscle cell plasmalemma.

_____

Skeletal muscle also has an extensive connective tissue component that, in addition to conducting blood vessels and nerves, becomes continuous with the connective tissue of its tendon. The tendon in turn is directly continuous with the connective tissue covering (the periosteum) of the adjacent bone. This connective tissue continuity from muscle to tendon to bone is the basis for movement of the musculoskeletal system.

What is the name of the loose areolar connective tissue covering of an individual muscle fiber?

_____

The perimysium is a collagenous connective tissue layer that groups several muscle fibers together into bundles called

_____.

Which connective tissue layer surrounds the entire muscle and merges with the connective tissue of tendons and aponeuroses?

_____

Nervous Tissue Review

From the PhysioEx main menu, select **Histology Review Supplement.** When the screen comes up, click **Select an Image Group.** From Group Listing, click **Nervous Tissue Slides.** To view slides without labels, click the **Labels Off** button at the bottom right of the monitor.

Nervous tissue is composed of nerve cells (neurons) and a variety of support cells.

Click slide 1.
Each nerve cell consists of a **cell body** (perikaryon) and one or more **cellular processes** (axon and dendrites) extending from it. The cell body contains the **nucleus,** which is typically pale-staining and round or spherical in shape, and the usual assortment of cytoplasmic organelles. Characteristically, the nucleus features a prominent **nucleolus** often described as resembling the pupil of a bird's eye (**"bird's eye"** or "owl's eye" nucleolus).

Click slide 2.
The cytoplasm of the cell body is most often granular in appearance due to the presence of darkly stained clumps of ribosomes and rough endoplasmic reticulum (**Nissl bodies/ Nissl substance**). Generally, a single axon arises from the **cell body** at a pale-staining region (axon hillock), devoid of Nissl bodies. The location and number of dendrites arising from the cell body varies greatly.

Axons and dendrites are grouped together in the peripheral nervous system (PNS) to form peripheral nerves.

What is the primary unit of function in nervous tissue?

_____

Name the pale-staining region of the cell body from which the axon arises.
.

_____

The support cells of the nervous system perform extremely important functions including support, protection, insulation, and maintenance and regulation of the microenvironment that surrounds the nerve cells.

Click slides 3, 4.
In the PNS, support cells surround both cell bodies (satellite cells) and individual **axons** and **dendrites** (Schwann cells). Schwann cells, in particular, are responsible for wrapping their cell membrane jelly-roll style around axons and dendrites to form an insulating sleeve called the **myelin sheath.**

Click slide 5.
Because Schwann cells are aligned in series and myelinate only a small segment of a single axon, small gaps occur between the myelin sheaths of adjacent contiguous Schwann cells. The gaps, called **nodes of Ranvier,** together with the insulating properties of **myelin,** enhance the speed of conduction of electrical impulses along the length of the axon. Different support cells and myelinating cells are present in the CNS.

What is the general name for all support cells within the CNS?

_____

Name the specific myelinating cell of the CNS.

_____

In the PNS, connective tissue also plays a role in providing support and organization. In fact, the composition and organization of the connective tissue investments of peripheral nerves are similar to those of skeletal muscle.

Click slide 3.
Each individual axon or dendrite is surrounded by a thin and delicate layer of loose connective tissue called the endoneurium (not shown.) The **perineurium,** a slightly thicker layer of loose connective tissue, groups many axons and dendrites together into bundles (fascicles). The outermost **epineurium** surrounds the entire nerve with a thick layer of dense irregular connective tissue, often infiltrated with adipose tissue, that conveys blood and lymphatic vessels to the nerve. There is no connective tissue component within the nervous tissue of the CNS.

What is the relationship of the endoneurium to the myelin sheath?

_____

_____

Endocrine Tissue Review

From the PhysioEx main menu, select **Histology Review Supplement.** When the screen comes up, click **Select an Image Group.** From Group Listing, click **Endocrine Tissue Slides.** To view slides without labels, click the **Labels Off** button at the bottom right of the monitor.

Thyroid Gland

The thyroid gland regulates metabolism by regulating the secretion of the hormones T_3 and T_4 (thyroxine) into the blood.

Click slide 1.
The gland is composed of fluid-filled **(colloid)** spheres, called **follicles,** formed by a simple epithelium that can be squamous to columnar depending upon the gland's activity. The colloid stored in the follicles is primarily composed of a glycoprotein (thyroglobulin) that is synthesized and secreted by the follicular cells. Under the influence of the pituitary gland, the follicular cells take up the colloid, convert it into T_3 and T_4, and secrete the T_3 and T_4 into an extensive capillary network. A second population of cells, parafollicular (C) cells (not shown), may be found scattered through the follicular epithelium, but often are present in the connective tissue between follicles. The pale-staining parafollicular cells secrete the protein hormone calcitonin.

Why is the thyroid gland considered to be an endocrine organ?

_____

What hormone secreted by the pituitary gland controls the synthesis and secretion of T_3 and T_4 (thyroxine)?

_____

What is the function of calcitonin?

_____

Ovary

The ovary is an organ that serves both an exocrine function in producing eggs (ova) and an endocrine function in secreting the hormones estrogen and progesterone.

Click slide 2.
Grossly, the ovary is divided into a peripherally located **cortex** in which the oocytes (precursors to the ovulated egg) develop, and a central **medulla** in which connective tissue surrounds blood vessels, lymphatic vessels and nerves. The oocytes, together with supporting cells (granulosa cells), form the **ovarian follicles** seen in the cortex at various stages of development.

Click slide 3.
As an individual oocyte grows, **granulosa cells** proliferate from a single layer of cuboidal cells that surround the oocyte to a multicellular layer that defines a fluid-filled spherical follicle. In a mature follicle (Graafian follicle) the **granulosa cells** are displaced to the periphery of the fluid-filled **antrum,** except for a thin rim of granulosa cells **(corona radiata)** that

encircles the **oocyte,** and a pedestal of granulosa cells (cumulus oophorus) that attaches the oocyte to the inner wall of the antrum.

Which cells of the ovarian follicle secrete estrogen?

_____

Uterus

Click slides 4, 5, 6.
The uterus is a hollow muscular organ with three major layers: the **endometrium, myometrium,** and either an adventitia or a serosa.

The middle, myometrial layer of the uterine wall is composed of several layers of smooth muscle oriented in different planes.

Click slide 6.
The innermost (nearest the lumen) endometrial layer is further divided functionally into a superficial functional layer **(stratum functionalis)** and a deep basal layer **(stratum basalis).**

Click slide 4.
A simple columnar **epithelium** with both ciliated and nonciliated cells lines the surface of the **endometrium.** The endometrial connective tissue features an abundance of tubular endometrial **glands** that extend from the base to the surface of the layer. During the proliferative phase of the menstrual cycle, shown here, the endometrium becomes thicker as the glands and blood vessels proliferate.

Click slide 5.
In the secretory phase, the **endometrium** and its **glands** and blood vessels are fully expanded.

Click slide 6.
In the menstrual phase, the glands and blood vessels degenerate as the functional layer of the endometrium sloughs away. The deep basal layer **(stratum basalis)** is not sloughed and will regenerate the endometrium during the next proliferative phase.

Which layer of the endometrium is shed during the menstrual phase of the menstrual cycle?

_____

What is the function of the deep basal layer (stratum basalis) of the endometrium?

_____

What composes a serosa?

_____

How does the serosa of the uterus, where present, differ from visceral peritoneum?

_____

Pancreas

The pancreas is both an endocrine and an exocrine gland.

Click slide 7.
The exocrine portion is characterized by glandular **secretory units** (acini) formed by a simple epithelium of triangular or pyramidal cells that encircle a small central lumen. The central lumen is the direct connection to the duct system that conveys the exocrine secretions out of the gland. Scattered among the exocrine secretory units are the pale-staining clusters of cells that compose the endocrine portion of the gland. The cells that form these clusters, called **islets of Langerhans** cells (pancreatic islets), secrete a number of hormones, including insulin and glucagon.

Do the islets of Langerhans cells secrete their hormones into the same duct system used by the exocrine secretory cells?

Cardiovascular Tissue Review

From the PhysioEx main menu, select **Histology Review Supplement.** When the screen comes up, click **Select an Image Group.** From Group Listing, click **Cardiovascular Tissue Slides.** To view slides without labels, click the **Labels Off** button at the bottom right of the monitor.

Heart

The heart is a four-chambered muscular pump. Although its wall can be divided into three distinct histological layers (endocardium, myocardium, and epicardium), the cardiac muscle of the myocardium composes the bulk of the heart wall.

Click slide 1.
Contractile **cardiac muscle cells** (myocytes, myofibers) have the same striated appearance as skeletal muscle, but are branched rather than cylindrical in shape and have one (occasionally two) **nucleus** (myonucleus) rather than many. The cytoplasmic **striations** represent the same organization of myofilaments (sarcomeres) and alignment of sarcomeres as in skeletal muscle, and the mechanism of contraction is the same. The **intercalated disk,** however, is a feature unique to cardiac muscle. The densely stained structure is a complex of intercellular junctions (desmosome, gap junction, fascia adherens) that structurally and functionally link cardiac muscle cells end to end.

A second population of cells in the myocardium composes the noncontractile intrinsic conduction system (nodal system). Although cardiac muscle is autorhythmic, meaning it has the ability to contract involuntarily in the absence of extrinsic innervation provided by the nervous system, it is the intrinsic conduction system that prescribes the rate and orderly sequence of contraction. Extrinsic innervation only modulates the inherent activity.

Click slide 2.
Of the various components of the noncontractile intrinsic conduction system, **Purkinje fibers** are the most readily observed histologically. They are particularly abundant in the ventricular myocardium and are recognized by their very pale-staining cytoplasm and larger diameter.

The connective tissue component of cardiac muscle is relatively sparse and lacks the organization present in skeletal muscle.

Which component of the intercalated disk is a strong intercellular junction that functions to keep cells from being pulled apart during contraction?

_____

What is a functional syncytium?

_____

_____

Which component of the intercalated disk is a junction that provides the intercellular communication required for the myocardium to perform as a functional syncytium?

_____

Blood Vessels

Blood vessels form a system of conduits through which life-sustaining blood is conveyed from the heart to all parts of the body and back to the heart again.

Click slide 3.
Generally, the wall of every vessel is described as being composed of three layers or *tunics.* The **tunica intima,** or *tunica interna,* a simple squamous endothelium and a small amount of subjacent loose connective tissue, is the innermost layer adjacent to the vessel lumen. Smooth muscle and elastin are the predominant constituents of the middle **tunica media,** and the outermost **tunica adventitia,** or *tunica externa,* is a connective tissue layer of variable thickness that provides support and transmits smaller blood and lymphatic vessels and nerves. The thickness of each tunic varies widely with location and function of the vessel. **Arteries,** subjected to considerable pressure fluctuations, have thicker walls overall, with the tunica media being thicker than the tunica adventitia. **Veins,** in contrast, are subjected to much lower pressures and have thinner walls overall, with the tunica adventitia often outsizing the tunica media. Because thin-walled veins conduct blood back to the heart against gravity, valves (not present in arteries) also are present at intervals to prevent backflow. In capillaries, where exchange occurs between the blood and tissues, the tunica intima alone composes the vessel wall.

The tunica media of the aorta would have a much greater proportion of what type of tissue than a small artery?

_____

In general, which vessel would have a larger lumen, an artery or its corresponding vein?

_____

Why would the tunica media and tunica adventitia not be present in a capillary?

_____

Respiratory Tissue Review

From the PhysioEx main menu, select **Histology Review Supplement.** When the screen comes up, click **Select an Image Group.** From Group Listing, click **Respiratory Tissue Slides.** To view slides without labels, click the **Labels Off** button at the bottom right of the monitor.

The respiratory system serves both to conduct oxygenated air deep into the lungs and to exchange oxygen and carbon dioxide between the air and the blood. The trachea, bronchi, and bronchioles are the part of the system of airways that conduct air into the lungs.

Click slide 2.
The trachea and bronchi are similar in morphology. Their lumens are lined by **pseudostratified columnar ciliated epithelium** with **goblet cells** (respiratory epithelium), underlain by a connective tissue **lamina propria** and a deeper connective tissue submucosa with coiled sero-mucous glands that open onto the surface lining of the airway lumen.

Click slide 1.
Deep to the submucosa are the **hyaline cartilage rings** that add structure to the wall of the airway and prevent its collapse. Peripheral to the cartilage is a connective tissue adventitia. The **sero-mucous glands** are also visible in this slide.

Click slide 3.
The bronchioles, in contrast, are much smaller in diameter with a continuous layer of **smooth muscle** in place of the cartilaginous reinforcements. A gradual decrease in the height of the epithelium to **simple columnar** also occurs as the bronchioles decrease in diameter. Distally the bronchioles give way to the respiratory bronchioles, alveolar ducts, alveolar sacs, and alveoli in which gas exchange occurs. In the respiratory bronchiole the epithelium becomes simple cuboidal and the continuous smooth muscle layer is interrupted at intervals by the presence of alveoli inserted into the bronchiolar wall.

Click slide 4.
Although some exchange occurs in the respiratory bronchiole, it is within the **alveoli** of the alveolar ducts and **sacs** that the preponderance of gas exchange transpires. Here the walls of the alveoli, devoid of smooth muscle, are reduced in thickness to the thinnest possible juxtaposition of simple squamous alveolar cell to simple squamous capillary endothelial cell.

What are the primary functions of the respiratory epithelium?

_____

_____

Why doesn't gas exchange occur in bronchi?

_____

What is the primary functional unit of the lung?

_____

The alveolar wall is very delicate and subject to collapse. Why is there no smooth muscle present in its wall for support?

_____

_____

What are the three basic components of the air-blood barrier?

_____

_____

Digestive Tissue Review

From the PhysioEx main menu, select **Histology Review Supplement.** When the screen comes up, click **Select an Image Group.** From Group Listing, click **Digestive Tissue Slides.** To view slides without labels, click the **Labels Off** button at the bottom right of the monitor.

Salivary Gland

The digestive process begins in the mouth with the physical breakdown of food by mastication. At the same time salivary gland secretions moisten the food and begin to hydrolyze carbohydrates. The saliva that enters the mouth is a mix of serous secretions and mucus (mucin) produced by the three major pairs of salivary glands.

Click slide 1.
The **secretory units** of the salivary tissue shown here are composed predominantly of clusters of pale-staining mucus-secreting cells. More darkly stained serous cells cluster to form a **demilune** (half moon) adjacent to the **lumen** and contribute a clear fluid to the salivary secretion. Salivary secretions flow to the mouth from the respective glands through a well-developed **duct** system.

Are salivary glands endocrine or exocrine glands?

_____

Which salivary secretion, mucous or serous, is more thin and watery in consistency?

_____

Esophagus

Through contractions of its muscular wall (peristalsis), the esophagus propels food from the mouth to the stomach. Four major layers are apparent when the wall of the esophagus is cut in transverse section:

Click slide 2.
1.   The **mucosa** layer adjacent to the lumen consists of a nonkeratinized **stratified squamous epithelium,** its immediately subjacent connective tissue (lamina propria) containing blood vessels, nerves, lymphatic vessels, and cells of the immune system, and a thin smooth muscle layer (muscularis mucosa) that forms the boundary between the mucosa and the submucosa. Because this slide is a low magnification view, it is not possible to discern all parts of the mucosa, nor the boundary between it and the submucosa.

2.   The **submucosa** is a layer of connective tissue of variable density, traversed by larger caliber vessels and nerves, that houses the mucus-secreting esophageal **glands** whose secretions protect the epithelium and further lubricate the passing food bolus.

3.   Much of the substance of the esophageal wall consists of both circumferentially and longitudinally oriented layers of muscle called the **muscularis externa.** The muscularis externa is composed of skeletal muscle nearest the mouth, smooth muscle nearest the stomach, and a mix of both skeletal and smooth muscle in between.

4.   The outermost layer of the esophagus is an adventitia for the portion of the esophagus in the thorax, and a serosa after the esophagus penetrates the diaphragm and enters the abdominal cavity.

Click slide 3.
Here we can see the abrupt change in epithelium at the gastroesophageal junction, where the **esophagus** becomes continuous with the **stomach.**

Briefly explain the difference between an adventitia and a serosa.

_____

_____

Stomach

The wall of the stomach has the same basic four-layered organization as that of the esophagus.

Click slide 4.
The **mucosa** of the stomach consists of a simple columnar epithelium, a thin connective tissue lamina propria, and a thin **muscularis mucosa.** The most significant feature of the stomach mucosa is that the epithelium invaginates deeply into the lamina propria to form superficial **gastric pits** and deeper **gastric glands.** Although the epithelium of the stomach is composed of a variety of cell types, each with a unique and important function, only three are mentioned here.

Click slide 5.
The **surface mucous cells** are simple columnar cells that line the **gastric pits** and secrete mucus continuously onto the surface of the epithelium. The large round pink- to red-stained **parietal cells** that secrete HCl line the upper half of the gastric glands, and more abundant in the lower half of the gastric glands are the chief cells (not shown), usually stained blue, that secrete pepsinogen (a precursor to pepsin).

Click slide 4 again.
The submucosa is similar to that of the esophagus, but without glands. The muscularis externa has the two typical circumferential and longitudinal layers of smooth muscle, plus an extra layer of smooth muscle oriented obliquely. The stomach's outermost layer is a serosa.

What is the function of the mucus secreted by surface mucous cells?

_____

Small Intestine

The key to understanding the histology of the small intestine lies in knowing that its major function is absorption. To that end, its absorptive surface area has been amplified greatly in the following ways:

1. The mucosa and submucosa are thrown into permanent folds (plicae circulares).

2. Fingerlike extensions of the lamina propria form **villi** (singular: **villus**) that protrude into the intestinal lumen *(click slide 7).*

3. The individual **simple columnar epithelial** cells (enterocytes) that cover the villi have **microvilli** (a **brush border**), tiny projections of apical plasma membrane to increase their absorptive surface area *(click slide 6).*

Click slide 7.
Although all three segments of the small intestine (duodenum, jejunum, and ileum) possess villi and tubular **crypts** of Lieberkühn that project deep into the mucosa between villi, some unique features are present in particular segments. For example, large mucous glands (**Brunner's glands**) are present in the submucosa of the duodenum. In addition, permanent aggregates of lymphatic tissue (**Peyer's patches**) are a unique characteristic of the ileum *(click slide 8).*

Aside from these specific features and the fact that the height of the villi vary from quite tall in the duodenum to fairly short in the terminal ileum, the overall morphology of mucosa, submucosa, muscularis externa, and serosa is quite similar in all three segments.

Why is it important for the duodenum to add large quantities of mucus (from Brunner's glands) to the partially digested food entering it from the stomach?

_____

Colon

Click slide 9.
The four-layered organization is maintained in the wall of the colon, but the colon has no villi, only **crypts** of Lieberkühn. Simple columnar epithelial cells (enterocytes with microvilli) are present to absorb water from the digested food mass, and the numbers of mucous **goblet cells** are increased substantially, especially toward the distal end of the colon.

Why is it important to have an abundance of mucous goblet cells in the colon?

_____

Liver

The functional tissue of the liver is organized into hexagonally shaped cylindrical lobules, each delineated by connective tissue.

Click slide 11.
Within the lobule, large rounded **hepatocytes** form linear cords that radiate peripherally from the center of the lobule at the **central vein** to the surrounding connective tissue. Blood **sinusoids** lined by simple squamous endothelial cells and darkly stained **phagocytic Kupffer cells** are interposed between cords of hepatocytes in the same radiating pattern.

Click slide 10.
Located in the surrounding connective tissue, roughly at the points of the hexagon where three lobules meet, is the **portal triad** (portal canal).

Click slide 12.
The three constituents of the portal triad include a branch of the **hepatic artery,** a branch of the hepatic **portal vein,** and a **bile duct.** Both the hepatic artery and portal vein empty their oxygen-rich blood and nutrient-rich blood, respectively, into the sinusoids. This blood mixes in the sinusoids and flows centrally in between and around the hepatocytes toward the central vein. Bile, produced by hepatocytes, is secreted into very small channels (bile canaliculi) and flows peripherally (away from the central vein) to the bile duct. Thus, the flow of blood is peripheral to central in a hepatic lobule, while the bile flow is central to peripheral.

What general type of cell is the phagocytic Kupffer cell?

_____

Blood in the portal vein flows directly from what organs?

_____

What is the function of bile in the digestive process?

_____

Pancreas

Click slide 13.
The exocrine portion of the pancreas synthesizes and secretes pancreatic enzymes. The individual exocrine **secretory unit,** or acinus, is formed by a group of pyramidal-shaped pancreatic acinar cells clustered around a central lumen into which they secrete their products. A system of pancreatic ducts then transports the enzymes to the duodenum where they are added to the lumen contents to further aid digestion. The groups of pale-staining cells are the endocrine **islets of Langerhans** (pancreatic islet) cells.

Renal Tissue Review

From the PhysioEx main menu, select **Histology Review Supplement.** When the screen comes up, click **Select an Image Group.** From Group Listing, click **Renal Tissue Slides.** To view slides without labels, click the **Labels Off** button at the bottom right of the monitor.

The many functions of the kidney include filtration, absorption, and secretion. The kidney filters the blood of metabolic wastes, water, and electrolytes, and reabsorbs most of the water and sodium ions filtered to regulate and maintain the body's fluid volume and electrolyte balance. The kidney also plays an endocrine role in secreting compounds that increase blood pressure and stimulate red blood cell production.

The uriniferous tubule is the functional unit of the kidney. It consists of two components: the nephron to filter and the collecting tubules and ducts to carry away the filtrate.

Click slide 1.
The nephron itself consists of the **renal corpuscle,** an intimate association of the glomerular capillaries **(glomerulus)** with the cup-shaped Bowman's capsule, and a single elongated renal tubule consisting of segments regionally and sequentially named the **proximal convoluted tubule (PCT),** the descending and ascending segments of the loop of Henle, and the distal convoluted tubule (DCT).

Click slide 2.
A closer look at the renal corpuscle shows both the simple squamous epithelium of the outer layer (parietal layer) of the **Bowman's capsule** (glomerular capsule), and the specialized inner layer (visceral layer) of **podocytes** that extend footlike processes to completely envelop the capillaries of the renal glomerulus. Processes of adjacent podocytes interdigitate with one another, leaving only small slits (filtration slits) between the processes through which fluid from the blood is filtered. The filtrate then flows into the **urinary space** that is directly continuous with the first segment of the renal tubule, the **PCT.** The PCT is lined by robust cuboidal cells equipped with microvilli to greatly increase the surface area of the side of the cell facing the lumen.

Click slide 3.
In the **loop of Henle,** lining cells are simple squamous to simple cuboidal. The DCT cells are also simple cuboidal but are usually much smaller than those of the PCT. The sparse distribution of microvilli, if present at all, on the cells of the DCT relates to their lesser role in absorption. The DCT is continuous directly with the **collecting tubules** and collecting ducts that drain the filtrate out of the kidney.

The large renal artery and its many subdivisions provide an abundant blood supply to the kidney. The smallest distal branches of the renal artery become the afferent arterioles that directly supply the capillaries of the glomerulus. In a unique situation, blood from the glomerular capillaries passes into the efferent arteriole rather than into a venule. The efferent arteriole then perfuses two more capillary beds, the peritubular capillary bed and vasa recta that provide nutrient blood to the kidney tissue itself, before ultimately draining into the renal venous system.

In which segment of the renal tubule does roughly 75–80% of reabsorption occur?

_____

How are proximal convoluted tubule (PCT) cells similar to enterocytes of the small intestine?

_____

What are the three layers through which the filtrate must pass starting from inside the glomerular capillary through to the urinary space?

_____

Under normal circumstances in a healthy individual, would red blood cells or any other cells be present in the renal filtrate?

_____

In addition to providing nutrients to the kidney tubules, what is one other function of the capillaries in the peritubular capillary bed?

_____